Working with disaster

Other titles from Longman

On Becoming a Manager in Social Work edited by Barbara Hearn, Giles Darvill and Beth Morris
Quality Assurance for Social Care Agencies by Emlyn Cassam and Himu Gupta
Making the Best Use of Consultants by Philip Hope
Getting Started with NVQ: Tackling the Integrated Care Awards by Barry Meteyard
Caring in the Community: A Networking Approach to Community Partnership by Steve Trevillion
NSPCC: *Child Sexual Abuse: Listening, Hearing and Validating the Experiences of Children* by Corinne Wattam, John Hughes and Harry Blagg
NSPCC: *Listening to Children: The Professional Response to Hearing the Abused Child* edited by Anne Bannister, Kevin Barrett, and Eileen Shearer
NSPCC: *From Hearing to Healing: Working with the Aftermath of Child Sexual Abuse* edited by Anne Bannister
NSPCC: *Making a Case in Child Protection* by Corinne Wattam
NSPCC: *Key Issues in Child Protection for Health Visitors and Nurses* edited by Jane Naish and Christopher Cloke
Making Sense of the Children Act (2nd edition) by Nick Allen
Coping with Violent Behaviour: A Handbook for Social Work Staff by Eric R. Brady

Social Services Training Manuals

First Line Management: Staff by Kevin Ford and Sarah Hargreaves
Effective Use of Teambuilding by Alan Dearling
Manual on Elder Abuse by Chris Phillipson and Simon Biggs
Developing Training Skills by Tim Pickles and Howie Armstrong
Training for Mental Health by Thurstine Basset and Elaine Burrel
Monitoring and Evaluation in the Social Services by David and Suzanne Thorpe

Working with disaster

*Social welfare interventions
during and after tragedy*

edited by Tim Newburn

Published by Longman Information and Reference, Longman Group UK Ltd, 6th Floor, Westgate House, The High, Harlow, Essex CM20 1YR, England and Associated Companies throughout the world.

© Longman Group UK Ltd 1993

Beautiful Boy
Words and music by John Lennon
© 1980 Lenono Music, administered by BMG Publishing Ltd for the United Kingdom and Eire
Used by permission
All Rights Reserved

You'll Never Walk Alone
(From Carousel)
Lyrics by Oscar Hammerstein II
Music by Richard Rodgers
Copyright © 1945 by Williamson Music
Copyright Renewed
International Copyright Secured
All Rights Reserved

All rights reserved. No part of this publication may be reproduced, stored in a retrieval system, or transmitted in any form or by any means, electronic, mechanical, photocopying, recording or otherwise, without the prior permission of the Copyright owner or a licence permitting restricted copying issued by the Copyright Licensing Agency Ltd., 90 Tottenham Court Road, London W1P 9HE.

A catalogue record for this book is available from The British Library

ISBN 0-582-10247-2

Typeset by The Midlands Book Typesetting Company
Printed and bound in Great Britain by
Biddles Ltd, Guildford and King's Lynn

For Mary

Contents

Introduction	Working with disaster Tim Newburn	1
1	We went to see Liverpool get to Wembley: the experience of a Hillsborough survivor Martin Hinks	9
2	The eye of the storm: police control of the Lockerbie Disaster Margaret Mitchell	22
3	Social Services after Hillsborough: managing the initial response David Mason	38
4	The Kegworth/M1 air crash: setting up a helpline and a support service David Whitham and Tim Newburn	51
5	Social work in the aftermath of the Zeebrugge Ferry Disaster Janet Johnston with Liz Beeson	65
6	Legal representation after the Kings Cross Fire Charles Pugh	78
7	The role of the general practitioner in the aftermath of the Lockerbie Disaster Margaret Mitchell	84
8	Mental health and social services: working together after Clapham Carolyn Selley	95
9	Responding to the needs of young people after Hungerford Elizabeth Capewell	105
10	Reaching out: running a staff care service in the aftermath of disaster Jane Harper	119
Conclusion	Social welfare after tragedy: what have we learnt? Tim Newburn	130
Appendix	Coping with a major personal crisis	140

Acknowledgements

My thanks first and foremost must go to the contributors to this book for making time in extremely busy schedules to write the pieces that appear here. Most of the chapters are highly personal in character and have taken considerable courage to produce. My hope is that their efforts will result in changed practices on the part of those organisations that respond to disasters. My work on the aftermath of the Hillsborough Disaster led me to believe that a book along these lines would be very valuable and I am most grateful to Alan Dearling for encouraging first the idea and then the project. Without his support, and that of Gerry Smale, Barbara Hearn and Nancy Dunlop at the National Institute for Social Work, the book would never have materialised.

Tim Newburn
London, January 1993

Introduction
Working with disaster

Tim Newburn

The 'disasters' literature is in some respects a relatively new and highly specialised one. The majority of academic work on this subject has been either North American or Australasian in origin, and the bulk of the work in this expanding area has either focused on disaster prevention and preparedness (civil defence and civil emergencies literature) or has concentrated on the medical and psychological impact of such events.[1] Until recently relatively little attention had been paid to the role and potential impact of social welfare agencies in the aftermath of such events.

This, as I have implied, is in the process of changing. The large number of disasters that occurred within a short period of time on the British mainland in the latter half of the 1980s focused attention on the extraordinary impact that such events have on the lives of ordinary people. It was impossible to ignore the long-term personal and social consequences that these fires, crushes and crashes had upon those unfortunate enough to be 'in the wrong place at the right time'. Because of this, the work of the welfare agencies that responded to these tragedies also came under some scrutiny, and plans or strategies for organising such responses have been produced by central government.[2]

There is, of course, still much to learn. Many agencies and authorities that would become heavily involved in difficult and protracted welfare work if there were to be a major disaster in their vicinity, have still not given much thought to what they would do under such circumstances, how they would do it, and what the consequences of so doing would be. Many others, although they have nominal plans in place, remain largely ill-prepared for such eventualities. I hardly need register that this is an unsatisfactory state of affairs, for there is an increasing academic, medical and welfare consensus that early, structured intervention may mitigate many of the worst long-term psychological and emotional consequences for those caught up in disaster. Furthermore, unprepared or ill-thought through actions by agencies responding to disasters not only carry the risk of exacerbating the distress being suffered by those they are supposed to be helping, but increase the chances that their own staff will themselves become 'victims' of the tragedy.

There are those that believe themselves to be well-prepared for responding to a civil emergency in their vicinity. Such authorities or organisations tend to have highly formulated and well and frequently tested plans. Some of these authorities will have been through one (or more) emergencies already and they and the others will, almost undoubtedly, have tried to gear their plans to what has been learnt from such experiences. Because of the catalogue of catastrophes in recent years, there is now much valuable experience that can be shared and built upon. The purpose of this book is to draw together some of this experience, but to do so in a way that is accessible and down-to-earth.

This book is, in at least one respect, quite different from all the other available material in this area, in that much of it is written in the first person. With two exceptions, all of the contributions here are from highly experienced professionals, all of whom have, in one capacity or another, been involved in organising or providing a service in the aftermath of one of the recent major tragedies in the UK. All of them have written personally about their involvement and their work, have looked critically at their experiences and have attempted to distil lessons that will be of general application to others faced with, or contemplating being faced with such work.

The reason that the material is presented in the first person is that this makes it more immediate and, consequently, easier to identify with. Disasters are frequently, if not generally, viewed as being highly unusual occurrences and, therefore, as being things that most of us are unlikely to experience. Whilst there is an element of truth in this, welfare agencies should not count on such good fortune. Those who have worked in the aftermath of disasters know only too well how poorly prepared they were for what they were eventually confronted with. Such is the power of these events that trained and experienced professionals frequently feel overwhelmed and imagine that they will not be able to cope with the task at hand. In my research on the 'care' response to the Hillsborough Disaster[3] I spent a great deal of time talking formally and informally to social workers and others about their experiences of responding to the disaster. One of their common complaints was that when they looked for written material that would help guide them in their work, they found relatively little that was of direct use. In particular, these workers said that what they would have valued especially were 'insider's accounts' of what doing the work was like. Indeed, this was borne out by the fact that the elements of training that they valued over all others were those parts that were given by people who had direct experience of disaster work, and which provided them with a first-hand account of elements of good practice as well as pointing out the potential pitfalls of the work.

The aim of this book, therefore, is to pass on some lessons from the experiences of a variety of 'disaster workers', whilst at the same time making explicit some of the greatest dangers of this highly stressful work, and doing so in a style that conveys a sense of the feeling, the emotion, and the circumstances of the aftermath of the terrible events that made the work necessary. It is important to state at this point that, in the main, these are the accounts of particular individuals. The experiences recounted herein, and the opinions expressed have no special claim to be 'the truth' – whatever that might be. Others might well provide quite different accounts. The purpose is not to produce either faithful historical accounts of the aftermaths of particular disasters, or dry quasi-scientific reports of organisational activity. Rather, these are accounts which describe in reasonable detail the intricacies of a number of aspects of post-disaster work, in a way that aims to give the reader some understanding of what it is like to be faced with such a situation. Armed with some feel for what others have experienced, workers will be better prepared for their own (potential) involvement in 'disaster work'.

As I suggested above, there is now a considerable academic literature exploring the emotional and psychological consequences of disaster. Equally, much has been written about bereavement and loss – though little of this literature has had disasters specifically in mind – and what has been written in relation to death and disaster reflects many of the concerns and received wisdoms of the more general bereavement literature. There is, by contrast, rather less known about the ways in which those who witness and experience disaster cope with having survived it, though there is an ever growing literature on what is commonly referred to as 'post-trauma stress' and its more serious, long-lasting or entrenched corollary, 'post-traumatic stress disorder'.

Whilst this work has resulted in an enormous advance in the understanding and treatment of those affected by sudden traumatic events, it is not designed to help welfare workers prepare for the kinds of tasks that they will be called upon to undertake. There is, in particular, a continuing public ignorance about 'survival', and it is working with survivors that social workers and others frequently find most problematic. In the opening chapter, Martin Hinks, who was present at Hillsborough, discusses the three years since his life was turned upside down by that event. In doing so he describes the range of feelings and emotions he has experienced during this time, and those things that he found helpful and unhelpful. Crucially, what Martin is able to do is to place the help he received from social workers within the context of all the other sources of help and support that he has called on since Hillsborough. It is clear from his experiences how important family, friends, colleagues and other survivors amongst others were, in helping him 'feel good about himself again'. Such a message is vital, for those providing support must not get into a position where they believe it is only they who can provide the support their 'client' needs. Martin's chapter forms the basis for the rest of the book.

There are two chapters that are not first-person accounts of working in the aftermath of disaster. Both of them (chapters 2 and 7) are, however, based on in-depth interviews with a large number of professionals (police officers and GPs) involved in the aftermath of Lockerbie. The first of these – that about the police – may at first sight appear out of place in a book about social welfare work. There are two reasons why this is not the case. First, without the organisational and controlling role played by the police after a major tragedy, much of the welfare work that is described in the following chapters simply could not take place. Secondly, the police are themselves frequently called upon to provide information and advice to those suffering trauma. It is important that social welfare agencies understand and appreciate the complexity of the role of this emergency service.

Organising and facilitating the work of others is also, at least in part, the subject of chapters three and four. In the first of these, David Mason describes the setting up of the social work service in the first week after the Hillsborough Disaster. As a Director of Social Services, he was one of those responsible for seeing that the major agencies and authorities worked together and, centrally, for representing these organisations in front of the world's media. Such a task might easily be underestimated, but he describes in graphic detail not only the occasional hostility and intrusiveness that has to be dealt with, but also the extent of the demands made upon spokespeople in the aftermath of disaster. In the early days this in itself appears to be a full-time job. It is also a crucial job for, as will be clear from the accounts throughout the book, the provision and control of information is of enormous significance both for the success of the organisation of the welfare response, but also to the welfare of those affected by disaster.

Chapter four also examines the work of a social services manager. In this case it focuses on David Whitham's role as the person responsible for putting together a service for the relatives of those injured in the Kegworth/M1 air crash. The injured were taken to three hospitals in the Midlands and David was the social work manager in Nottingham's major hospital which, because it has a neurosurgical unit, took a large number of cases. He describes the intricacies of setting up a helpline in the hospital, and an area where relatives of the injured could be cared for or could simply find some peace and quiet. Like David Mason, he is able to convey the urgency with which decisions have to be taken, and he provides a checklist of items based on his experience which will not only help any managers involved in a disaster response, but also any staff who may be called upon to offer their services.

Janet Johnson, on the other hand, although herself occupying a managerial role, was also working on the 'frontline' with survivors from the Zeebrugge Ferry Disaster. Consequently, she is not only able to recount how a social work service was put together in the aftermath of that tragedy, but also to describe the nature of the problems that survivors presented, together with details of some of the work undertaken with them. Her chapter concludes with a description of the process of psychological debriefing. This process is of potential benefit not only to the 'primary victims' of the disaster, but also to the staff involved in the care work in the aftermath. One of the issues raised in this chapter, and also in many others, is the potential impact of the work on staff. This is frequently ignored in texts on disaster and disaster work, but comes through strongly whenever the experiences of staff are examined directly.

As both David Mason and David Whitham point out, the helpline is designed to be both a source of information and help, and a central point at which offers of help and information can be co-ordinated. In the chaos that characterises the immediate aftermath of disaster, it is information more than anything else that most people need. This remains true in some senses even after much time has passed. For example, disasters tend to give rise to a number of legal proceedings, including official inquiries, court cases and inquests. Unfortunately, as Charles Pugh outlines on the basis of his experience as a lawyer representing some of the Kings Cross families, such proceedings are rarely organised in such a way as to provide the information that the bereaved and survivors so desperately need. Furthermore, he argues that, in this respect, lawyers have a crucial welfare role to play, for they, in conjunction with social workers or other counsellors, can begin to redress this problem by providing full and detailed accounts of the disaster for the relatives of those who died. In conclusion, he makes a number of recommendations about how legal proceedings should be conducted in the future, and what training lawyers require in order to undertake such welfare work.

The following two chapters both focus on aspects of the health service. Margaret Mitchell analyses the role of GPs in the aftermath of Lockerbie and shows how they are a potential interface between those affected by disaster and the wide variety of other services that are available. Local doctors in an afflicted community have a much broader role than simply responding to people's immediate physical problems. They need to look beyond the primary presenting problems for signs of the emotional and psychological trauma that often accompanies experience of such events. By the same token, given the central role of the GP, there is much that other agencies have to learn about organising proper links and ensuring that the other services that are available are not only known, but are understood by local doctors.

Carolyn Selley examines the reality of liaison between health and social services in the aftermath of the Clapham rail crash. Sadly, but perhaps predictably, she chronicles a variety of problems she faced in attempting to ensure that there was consistent communication between agencies. Nevertheless, the concluding message is the optimistic one that social workers and health service based psychotherapists and psychiatrists have much to learn from each other and, in tandem, much to offer those affected by disaster.

Elizabeth Capewell also looks at co-operation with social services, but her viewpoint is that of a Youth and Community Officer in a local education authority. She focuses on a subject that is all too frequently sidelined or even completely ignored: the needs of children. In addition to detailing the impact that the Hungerford shootings had on some of the children she worked with, she explains just how difficult it was to set up a service for them in the face of the open resistance she encountered from many adults who adopted a 'least said, soonest mended' attitude to children's feelings. In attempting to remedy this situation, she provides some guidelines for working with children and within schools. She also describes in honest terms just how profound an impact her experiences after the shootings had on her life. As she says 'Five years later I can only now face and feel the pain'.

It is precisely such an impact, an impact that all the contributors make reference to, that makes it so vital that 'staff care' is taken seriously as an issue and not, as is more usually the case, simply paid lip service to. In chapter ten, Jane Harper describes the staff support service that was set up after Hillsborough of which, for two years, she was the co-ordinator. This was the first and only time such an initiative has been run in the aftermath of a disaster in the UK, and she provides a unique insight into the problems, pitfalls and positive lessons that were learnt from this experiment. Not only is she able to confirm what the workers in the previous chapters say about the powerful impact of disaster on all those that come into contact with such an event, but she also shows how the general methods that social workers adopt to work with the bereaved and survivors are also the most appropriate for a staff care service as well.

The broad range of experiences reported here provide a general overview of the variety of services and organisations that may become involved in the welfare response to a major tragedy. Despite the large number of recent disasters, there remains widespread ignorance about the reality of such events and their aftermaths and, especially, what is involved in providing a service for the 'victims' of the event. The primary aim of this volume is to convey some sense of the reality of the aftermath of disaster, whilst at the same time detailing some of the most basic (and therefore important) nuts and bolts of the work of 'welfare' agencies and their staff.

Notes

1 See for example the bibliography produced by Bradford University's Centre for Disaster Research.
2 A Working Party sponsored by the Department of Health, and run by CRUSE: Bereavement Care produced a report in 1991: *Disasters: Planning for a Caring Response*, HMSO; more recently, the Home Secretary's special advisor, David Brooke, has produced guidance on civil emergencies: *Dealing with Disaster*, HMSO.
3 Newburn, T. (1993) *Making a Difference? Social Work after Hillsborough*. London: National Institute for Social Work.

1 'We went to see Liverpool get to Wembley': the experience of a Hillsborough survivor

Martin Hinks

> **Martin Hinks**, *a Registered Mental Nurse with a Certificate in Child and Adolescent Mental Health and a Diploma in Management, is an experienced Mental Health Nurse. He is also a Liverpool supporter who was at the Hillsborough stadium on 15th April 1989. He is currently working outside the National Health Service and living in Shetland. Anyone wishing to contact Martin should write to him via Tim Newburn (PSI, 100 Park Village East, London NW1). This chapter reports some of Martin's experiences since Hillsborough, the support he received from family, friends, colleagues and professionals, together with a guide to those things that he found helpful during his process of recovery.*

> Life is what happens to you, while
> you're busy making other plans!
> *Words and Music by John Lennon, 1980*

Nobody could have foreseen how prophetic these words from one of John Lennon's final songs would be. Not long after, he was gunned down in a New York street outside the apparent safety of his own home. The sentiments mean something more to me now, having experienced first hand the mind-blowing and unexpected happenings of 15th April 1989, when 95[1] football supporters died in an unnecessary and avoidable catalogue of tragedies. The aim of the 95, as mine had been, was simply to see their favourite team win a game of football. It was a day out. We didn't think our lives would be threatened. It is this invasion of the concept of safety, and the ensuing feelings of vulnerability, that has caused me most difficulty and also many others I have spoken to since.

I write three years after the tragedy, when I suppose I had imagined myself to be 'okay', 'normal' again, or at least 'over the worst'. It is important at this stage to begin to question the commonly held and even more commonly stated myth; that 'time heals'. Time alone is unlikely to do very much to heal – not without the many other facets of what often proves to be a very active and unpredictable process. My case is no exception.

One thing is certain. Walking away physically unscathed did not make me feel 'lucky', and I believe it can only ever be the start of the process of coming to terms with what I had experienced. As an experienced mental health nurse, I have found myself knowing what I wanted to do, what I 'should' do, how I'd have preferred to feel, whilst at the same time finding it extremely difficult, sometimes impossible, to put these lessons into practice.

Early days

I have a very blurred memory as far as driving from Sheffield to my home in the Midlands goes. I can remember bits and pieces; like being on the hard shoulder of the motorway in tears, and stopping off at a café in a town miles from my route. I can remember very clearly the feeling of euphoria at pulling up hours later on our front drive and being held firmly by my wife who had been following it all on TV. We cried together for a few moments, and then all I remember is finding myself being approached by a friend in a nearby park. 'I suppose it's a silly question' he said, 'but how are you feeling?'

It was far from silly. It was the first recognition apart from my wife's that I was suffering or, to use one of my own terms, that 'my head was gone'. I imagine my reply would have been along the lines of 'I'm okay'. How often must I have said this during the ensuing months? After all – what else can you say? The following 24 hours continued in this vein; okay one minute, distant the next. Conscious of my behaviour and feelings one minute; then lapsing into what people have since told me were dream-like states. 'Martin' they'd say, 'You're not all there'.

The following day, after what I can only describe as a sleepless night, I felt a burning need to do something. Perhaps I should return to the scene, or pray to God for the families of the dead, or drive and drive and drive. I couldn't keep still, and neither was my mind able to focus on anything other than the previous day's horrific events. It was impossible to face the day trying to be 'normal'; Sunday roast, church, TV and tea and sandwiches. I was persuaded not to return to the ground, a wise decision at that early stage, I think. Instead, my family and I travelled to Liverpool and attended the service at the Cathedral. The sight of so many people armed with flowers, wreathes, scarves and other items to express their grief was one that will live with me forever. I began to realise that this had hit a vast amount of people from all over the country. I also began to accept the fact that I had actually been there and had been powerless to do anything very much to help. It was a monstrous feeling.

I mistakenly tried to work, as normal, on the Monday. I broke down, couldn't concentrate. I felt I needed to run away – but where to? I was most fortunate to have very sensitive and supportive colleagues and managers. That afternoon, having been encouraged to watch a recording of the TV coverage of the event with my boss, he and another friend accompanied me, once again, to Anfield, home of Liverpool Football Club. The ground had been opened for the public in response to the hundreds who had congregated outside, tearful, dazed, walking aimlessly and needing to find an outlet for expression of their feelings. I laid down my scarf on the Kop, where I used to stand. An old teddy bear of mine found a place in the goal area, along with some red and white flowers and an old Liverpool football shirt. The decision by the club to open the ground in this way was indeed a wise and sensitive one. In the following week I made more trips up the motorway to Liverpool, where I felt surrounded by people who understood; who were feeling the same way. People descended on the city from everywhere, including Ireland, Scotland and parts of Europe. Again, the growing realisation that I had endured and witnessed a disaster of major proportions was paramount. Also the feeling that I was not alone.

The first week was dominated by feelings of intense pain, sadness, anger, helplessness and loss. I had a constant lump in my throat and I had already begun to distance myself from certain people in my life, at work as well as socially. I also felt a distinct feeling of emptiness, and everything seemed pointless. The support of close friends, their actions, expression of grief and of other emotions was sensitive, supportive and very important to me. But to be honest, I needed to be guided more often than not, listened to a lot, and 'held' safely, if not physically then just not being left alone was important. My wife and kids were already beginning to feel the effects, because the disaster had very quickly become an all-consuming part of my existence. I wanted to be happier, more enthusiastic, but *I couldn't*. The added flashbacks and constantly ruminating on the events simply made it impossible.

The following months

I realised by the end of the first week that I was going to need some help to come to terms with it all. I did not seek professional help, but started to talk and talk and talk about it all to anybody who would listen. Although at times it did help, I felt a burden. I began to see myself as a boring individual, at least to these people, who with the best will in the world seemed unable to see beyond 'it's over now, you're lucky to be alive'. At the time these responses felt about as comforting and acceptable to me as 'pull yourself together' or 'life goes on'. Part of my life had, in fact, stood still, and I found myself unable to switch off. I was constantly trying to make sense of it all, asking why? and how? Of course there was no sense to it, but I needed to find this out over time, and then accept it. This is not easy to do.

Writing poetry seemed easier for me. It came in an instant, some angry, some sad, some making sense and some not. The beginnings of a scrap-book were begun, and to this day I still refer to it from time to time, especially at the nearing of the anniversary. For me, writing and talking about it came naturally, and has proved useful, though common sense tells me that not everybody will be quite so comfortable with this form of coping strategy. I avoided close social contacts for many months. This was primarily because I felt vulnerable and insecure. How could I be safe when, within the space of a few months, we had been traumatically burgled, family members had been very ill, personal and close loss had occurred, and then – Hillsborough! Even without these other things, the horrific nature of the events at Hillsborough would have had a marked effect on anybody.

The feelings of uselessness and poor self-worth were compounded by the fact that I was absolutely convinced that nobody could understand what I was going through. I began to be afraid of crowds, and my concentration became poor. I felt that the high standards I had set in my work performance were suffering. My ability to concentrate for long periods was disturbed. At times, my judgement and assessment of situations became somewhat blurred. Above all, I felt I was becoming one of the most irritable human beings imaginable. I found myself acting out of character in a number of small ways, and I was plagued by feelings of guilt at not having been able to save any lives at the stadium. The guilt was also partly related to the fact that I was still alive, and there was also a spin-off guilt resulting from allowing these feelings to affect my family so significantly that there were times when my wife and daughters must have felt 'husband and father-less'. At times I found myself, in a way, 'outside myself', looking in. I rarely liked what I saw. I was getting angry, but displacing it onto other people, or more usually through activities such as squash, football, cricket and running. To this day I have some doubts that I was *always* able to *appropriately* channel or displace my feelings of anger and confusion.

Bit by bit, events of the day were unfolding – at times all at once and it felt like too much, and at other times only small things at any one time. Never, though, did there seem a day when nothing new was coming to mind, or when Hillsborough wasn't with me. In many ways this is still the case, although it is a lot less painful now, and I find myself for the most part able to discuss it or write about it in a less emotive way. During the first five months or so there was a pervasive feeling of duties to be performed, of 'unfinished business'. The drive to do something, to be actively involved with others affected by the tragedy, was powerful and still is. For twelve months afterwards I kept reminding myself that people are constantly faced with potentially stressful and emotional situations, many worse than me, and that no two individuals will respond in exactly the same way. Also, that the vast majority survive and eventually learn to cope. Telling myself these things, that had so often rolled off my tongue in my attempts to help others rebuild their shattered lives, had little positive impact on me. This discovery makes me question our counselling strategies.

I just could not see an end to the turmoil, and I soon realised that after nearly 12 months I had not spoken about the tragedy to anyone who had been there. Neither had I spoken to a professional despite my awareness that all was not well. In hindsight I wonder if by leaving it so long I hindered my own recovery. I may never know. What I do know is that a massive turning point was about to take place.

Twelve months after the tragedy

I eventually felt able to engage the help of professionals specialising in disaster work. The spur was the growing realisation that I was unable to regain control all by myself. My close relationships were showing signs of stress and tension as a result. The trigger incident which led me to make the first contact was the utter panic I felt one particular day which led me to run and run, tears in my eyes, to avoid nothing more sinister than a set of school railings which had somehow reminded me of the bars that had hemmed in the Hillsborough crowd. After a few meetings with a professional social worker, where the process of order, perspective and gradually working through the turmoil truly began, we attended a memorial service together, exactly 12 months after the tragedy. At the same time I found the Hillsborough Centre, an old sandstone lodge situated in the grounds of Stanley Park, little more than a stone's throw away from Anfield, home of Liverpool FC. As I entered the centre the togetherness and warmth was an instant tonic, and I vowed to make more use of the place from then on.

Staffed by experienced social workers and counsellors, the centre facilitated self help groups for survivors, the bereaved, as well as being a central contact point for anyone affected in any way by the tragedy. The Hillsborough Centre became a lifeline for many, including myself. I attended the centre on a weekly basis for about three months, during which time I was able to share some of the difficulties I had been experiencing with others who were going through similar situations. Listening to others' descriptions of events, and ways in which they were coming to terms with it all gave me the courage to carry on – encouraged me to see an end to it all. In the groups we were able to help each other to feel less 'freakish', more normal, more hopeful of positive outcome. Feelings of anger, guilt, pain, confusion, helplessness, fear and anxiety, to name but a few, were allowed to surface. There were no negative judgements, no pressure for a 'quick cure'. Instead there was a feeling of togetherness, of understanding and of tolerance. If I didn't want to speak on a deep, personal level on any particular day, I didn't have to. Whatever was wanted or needed was perfectly admissible. I saw others who were going through a more difficult time than myself, some had not been able to return to work and some had ended their relationships as a result of the total chaos that their lives had become. I found myself beginning to put it all in context, into some kind of order. Subsequently I have been able to utilise the offer of a one-to-one counselling relationship during which time being able to watch, with supervision, video replays of the events proved most useful. I was able to retrace my steps on the day and put the finishing touches to the order and the gaining of perspective. I have, through the centre, been able to slow down my desire for a quick return to 'normality' and replace it with a more realistic, measured, supported build-up towards a 'new norm' within which a stronger and wiser person exists.

Sharing has undoubtedly been crucial. The Hillsborough Centre offered me the chance to talk about it all with those who understood. I had, before this, been wanting to blame every negative aspect in my life, every lump in my throat, on Hillsborough. Talking and listening to others enabled me to begin to separate what pain belonged where, and this became a central feature of my recovery. I began to apply more positive associations to the tragedy. I began to seek strategies to ensure nothing so destructive ever happened again. I began to ask myself what I could do to contribute to making sure lessons learnt were properly shared. Talking to others who had suffered as a result of Hillsborough and other disasters did much to alleviate excessive inward looking and obsessive thinking on my part. Only by people writing and talking openly about the after effects of such things can we begin to become more understanding and responsive. The fear of being seen as 'weak', or 'not coping', or even 'attention seeking' can at times prevent this from happening.

In my case, the commitment to turn the years of turmoil into a contribution to more informed future responses has overridden these fears. Hence my willingness to describe my experiences via articles and writings such as these. You see, the fact that 95 died, over 400 received hospital treatment and around 700 were physically injured, cannot and does not, record the full extent of the Hillsborough disaster and its many victims. It all goes far deeper than that, and it is crucial that those in the helping professions realise the complexities of post-traumatic reactions. The need to talk and to express is paramount. Avoiding the danger of it all becoming self-perpetuating, and thus making matters worse, requires an incredible degree of skill and patience from all concerned. It is not enough just to care, or to care alone, although it is obviously an integral part, or beginning, of a helping relationship. But there can never be a tangible date or time-span after such experiences when the sufferer should be expected to feel 'wholly' well again. My own attempts to recover following the Hillsborough disaster were in many ways hampered by feelings and symptoms closely associated with very normal and understandable processes of coping again after a traumatic experience. Just allowing myself the permission to say that is a sign of movement and progress, believe me.

To this day, although I am largely beginning to feel good about myself again, and stronger, I am often aware that the events of 15th April 1989 and its aftermath are rarely all that far from the surface. Nobody should underestimate the levels and depths of the 'black hole' which hits people after such events. For the survivor, it hurts, it dominates, it blows your mind. It creates all kinds of

Figure 1.1 Key themes central to recovery (for me)

1. Supportive family, friends and work-mates.

2. Time out from pressures, demands and the pressure to be 'normal' (whatever that is!)

3. Active expression of grief and pain – not only in talking but also by writing, laying down of flowers and LFC memorabilia, returning to the scene (with support) etc. . . .

4. Counselling – to help put it all in context and to help to separate what pain belongs to the disaster and my role in it, and what was pertinent to other aspects of my life.

5. Being with people who had been through it, or something similar, who therefore have an acute understanding.

6. Working through different stages of recovery. Pacing it. Not all at once.

7. Self-help groups. 'So I'm not alone after all!'

8. Talking and listening to bereaved families and other survivors.

9. Being helpful to others who are suffering without imposing or missing essential 'work' for yourself.

10. Writing articles for nursing/professional journals to begin to 'spread the word'.

11. Leaflets and research being made available.

12. Sports and activity (e.g. cricket, football, badminton, squash, swimming, walking).

13. Some time alone, in privacy, collecting thoughts.

14. Measured with some time with others. Early on, being left alone caused me immense fear and difficulty.

15. Keeping a scrap-book.

16. A 'link' magazine, keeping in touch with others.

17. Setting up a local self-help group (I lived 100 miles or so from Liverpool).

18. Crying! In 'appropriate' and 'safe' places/times.

19. Getting angry. In safe/controlled environments.

20. The healing balm of time – but not without the other things!!

21. Rest and relaxation. (A place to retreat to every now and then is indeed an added bonus!)

22. Professional help and counselling.

23. *Music:* compiling tapes, playing music, enjoying listening to music.

24. Poetry and prose, both writing and reading.

25. The Hillsborough Centre – knowing that there's always someone close at hand even if only at the end of a telephone.

26. Friendships. Social life. As much as possible to maintain good, fun aspects of your life.

27. Lots of physical demonstrations of love, affection, hugs, cuddles, etc. It reaffirms that you belong and someone cares.

28. Open communication channels – a means to an end.

29. The gradual installation of hope for a positive outcome on a personal level.

30. Helping to ensure lessons learnt are properly shared.

18   Working with disaster

Figure 1.2 What did it feel like?

The following are key words which will hopefully give some summary of how the aftermath of being a survivor affected me, at different times and to differing degrees.

Shock	
Numbness	
	Need to return to the scene
Loss of control	
Fear	
Pain	
	Seeking someone to blame
Extreme sadness	
Anger	
	Need to meet others who were there
Guilt and self-blame	
Depressive thoughts	
Anxiety and startled responses	'No Way Out'
Irritability	'Turmoil won't end'
Relationship difficulties	'I should have died too'
Tension	
Flashbacks	
Nightmares	
Acting out of character	'I have nothing to look forward to'
Behaviour changes	
Constant replay/unable to switch off	
Loss/emptiness	
Waking up at night	
Unable to get off to sleep	

Need of people around me	
Alternatively, a desperate need to be alone with my thoughts	
Feelings of helplessness	
Poor concentration	What right have I to be feeling this way? I survived it after all!
Changes in eating habits	
Changes in drinking habits	
Going over the event in my mind	Am I going mad?
Trying to make sense of it all	
Putting it into context	Things that didn't used to bother me beginning to seem like 'the end', too much'
Worry/excessive anxiety	
Motivational changes	
Overwhelmed	
Poor self image	
Loss of confidence	
Feelings of panic	
Pessimistic thoughts	
Need to cry a lot	
	The need to accept what has happened as part of a new me – and to move on
Need to talk a lot	
Need to write about it	
Feelings of powerlessness	
Feelings of vulnerability	
Fear of recurrence	
Feelings that can only be described as total and utter despair	

imbalances and shatters your whole world. You feel as though you're losing everything that you value – at least at times I did. Seeking to put it all 'out of mind' is a mistake, whilst striving to put it into context must be a more sensible and realistic target. To set a time limit, I stress, is folly indeed. Nevertheless, allowing things to drift without utilising help offered, would also be asking for trouble. My heart goes out to anyone suffering after major experiences of loss and pain. I feel a heightened sense of others' pain and suffering, and I hope I can be a better person as a result. If not, then what was the point of it all? The individual may well take solace and comfort from being regularly reminded that he or she is still intact, still the same person as before, but with 'new bits' added. These 'bits' could strengthen the person concerned and others too, in time.

As Oscar Wilde wrote in *Ballad of Reading Gaol*, 'nothing in the world is meaningless, suffering least of all.' I would like to finish with some words from a very famous song.

> At the end of the storm,
> There's a golden sky,
> And the Sweet Silver Song of the Lark . . .
> Walk on through the wind,
> Walk on through the rain . . . .
> Walk on, Walk on,
> With hope in your heart,
> And you'll never walk alone.
> You'll Never Walk Alone
> (From *Carousel*)

It is so important not to feel alone when faced with the major life imbalance thrown up by a disaster. Often, to get to 'the end of the storm' requires much soul-searching and pain to be worked through. This *cannot* be accomplished alone. The desire to forget, whilst natural, is unhelpful. Feeling that you belong and that help is available rekindles hope for the future. I hope that my contribution will be of some help. Thanks for listening . . . and good luck!

Useful reading

I have found a number of books and journals most effective in working through the confusion and chaos. Among these are a number of leaflets and articles published by CRUSE-Bereavement Care, 126 Sheen Road, Richmond, Surrey.

The following are texts I would advocate most strongly, to anybody who wants to know more about disaster, survival and post-trauma reactions.

Gibson, M. (1991) *Order from Chaos*. Venture Press.
Kushner, H.S. (1982) *When Bad Things Happen to Good People*, Pan Books.
Murgatroyd, S. and Woolfe, K. (1982) *Coping with Crisis*. Harper and Row.
Muss, D. (1991) *The Trauma Trap*. Doubleday.
Parkes, C.M. (1988) *Bereavement: Studies of Grief in Adult Life*. Penguin.
Raphael, B. (1986) *When Disaster Strikes*. Hutchinson.
Rogers, C.R. (1980) *A Way of Being*. Boston: Houghton Mifflin.
Stewart M. and Hodgkinson, P. (1991) *Coping with Catastrophe*. Routledge.

Note

1 Since this was written the tragic case involving the Bland family has been decided in the courts. The death of Tony Bland brings the number of deaths as a result of Hillsborough to ninety six.

2 The eye of the storm: police control of the Lockerbie Disaster

Margaret Mitchell[1]

Dr Margaret Mitchell is a Reader in Psychology at Glasgow Caledonian University. She is a chartered psychologist and a member of the International Society for Traumatic Stress Studies, and has studied the effects of disaster on several groups including the police and other emergency personnel. In examining the work of the police, this chapter concentrates on the organisational and controlling role they play in the aftermath of disaster, and looks at the way in which they facilitate the work of others.

The following chapter is based on interviews with senior and other ranks of police officers involved in the immediate aftermath of the Lockerbie air crash.

The disaster

An event of enormous dimensions took place in Lockerbie on the evening of the 21st of December 1988. It mobilised emergency and volunteer services from all over Britain, and changed this small rural community for ever. At 18.25 hours on the 21st of December 1988 a Boeing 747 aircraft on Pan American Airways Flight 103 took off from London, bound for New York. On board were 243 passengers and 16 flight crew. At 19.03 the aircraft was flying at 31,000 feet above south west Scotland when an explosion occurred on board causing the break-up of the plane.

The circumstances of the Lockerbie disaster were unique from many perspectives. It impacted on a small community in southwest Scotland, but involved people of many different nationalities. The physical and geographical scale of the disaster was unique – pieces of the aircraft, the cargo and all the occupants of the plane were strewn over an area of 845 square miles. The fact that so few were injured, and that it was a deliberate act of murder rather than an accident also made it unusual. Investigations to date have shown that all 259 persons on board the aircraft died, and eleven local residents were killed as a result of aircraft wreckage falling to the ground. Of these, the remains of 253 victims have been identified while 17 are missing presumed dead and who in all probability will never be found. An area of 845 square miles has already been thoroughly searched and searches of particular areas will continue as necessary. The impact of the disaster continues. It will never be over for the bereaved relatives; some emergency workers who assisted at the site and some local residents continue to be profoundly affected by the trauma of the disaster. Apportioning blame for this tragedy has not yet been resolved and, despite the protracted police inquiry, the named suspects have yet to be brought to justice.

The night of the disaster the residents of Lockerbie, sitting at home that evening just a few days before Christmas, heard a roar and a whine which they report sounded 'like bombs being dropped'. This was followed by an impact which was recorded at the Eskdale Seismic Earthquake Centre. When the plane's fuel tank hit the ground a fireball erupted which was clearly visible for miles around. People left their homes to see what had happened. In the town, crowds of people were walking in the direction of the flames, some with babies in push-chairs and some cycling. There were puddles of burning kerosene, and debris lying everywhere in gardens, on the roads and pavements, on the roofs of the houses. On the ground bodies or body parts were also quite visible.

'It was obviously a major catastrophe, although at the outset it was not known what precisely had happened. For the first three to four hours after the impact, there was absolute chaos in the town and stories about the possible cause and effects of the catastrophe travelled fast.' The petrol station was thought to be in imminent danger of exploding, houses were on fire and some were reported to have disappeared completely. Sherwood Crescent had been turned into a huge flaming crater (where the main fuel tank of the aircraft had landed and exploded). Tons of debris and earth was thrown on to the adjacent main Glasgow to Carlisle motorway and there were reports of several cars on fire. An aircraft engine had landed and fractured the main water-pipe, cutting off the supply for the fire service. There were problems with basic amenities: there was no light to search the houses for the dead and injured because the electricity lines had been cut and a gas pipeline had been severed, presenting further danger.

'People were more shocked than panicked or frightened and they all asked the same questions. What is it? What can we do?

They were coming across parts of bodies and one resident asked what she was to do with an arm found in her front garden. Another woman asked if I had seen her mother – it later transpired that her mother was one of the elderly residents of Sherwood Crescent who died in the tragedy. The main passenger fuselage landed in the Rosebank area, a council housing scheme to the east of Lockerbie, by some miracle between two rows of houses. In all this it is quite amazing that a great many more people on the ground were not killed: five people were injured, two suffered serious burns and three were treated for shock and minor injury. Another elderly lady escaped with her dog and her canary from the end house at the Crescent where the fuselage had torn down the gable end. She reported sitting in her house when the wall "just disappeared"'.

'The scene in the town was eerie – to say the least. There was no light, no power, very little sound, only the rustle of the wind, flickering torches, bodies and mutilated parts of bodies, and debris lying around. It was so quiet it was almost like watching a film. There was a smell of burning kerosene and smoke was rising in little pyres all over the place like some futuristic scene. Helicopters were flying overhead. You were aware of people talking in small groups but their voices were sort of muffled by this strange eerie quiet. Even though they were quite close they seemed far away.'

The immediate response

In this catastrophic situation it is only through the police taking control that the emergency services and other necessary welfare support can function. Specifically, the role of the police in this immediate aftermath is to facilitate the preservation of life by ensuring access for other emergency services. It is also important to establish as quickly as possible the precise nature of the disaster and to assess the appropriate emergency response. Until this is known nobody is in a position to dictate what is required of other agencies. Full emergency medical procedures were implemented at the nearest hospitals until midnight when, four hours after the impact, it was obvious that there were no survivors, at which time the medical alert was called off.

The first reaction of householders in an emergency situation of this type is to telephone, with the result that the lines quickly become overloaded and inoperable. The emergency services were inundated by calls, although by midnight we (Strathclyde Police) had set up a Casualty Bureau at the Force's Headquarters in Glasgow to log telephone calls from worried relatives and to initiate enquiries into missing persons by officers in Lockerbie. Initially there were great difficulties accounting for everyone. The fate of the drivers of the burned-out cars on the A74 was particularly perplexing, and it was not until some time later that it was discovered that in each case they had walked free of the car before it had burned.'

'Telephone or radio communication into and out of the area was an immediate and pressing problem. Emergency personnel were unable to make contact, with the result that emergency services from neighbouring areas just kept on arriving. To alleviate the immediate communication problems, British Telecom lifted and re-directed fifty domestic telephone lines, and although this helped, it had a "knock-on effect" for the police. The son in Scarborough, or London, or Shetlands, or Orkney, or Glasgow, or wherever, whose mother, father, uncle, brother stayed in Lockerbie telephoned to see if they were all right. Because the lines were down or jammed they didn't know if their relative was alive, dead, injured or what. Naturally, they telephoned the police to get information thereby jamming those lines too. To be honest the lifting of the domestic lines actually caused that particular problem. We learned as we went along.'

That night in accordance with the Regional Emergency Plan all the relevant departments (social work, education, roads and transportation, property services, information technology, communications, public relations, police) met in the Regional Council Offices in Dumfries, the main local town and some 20 miles distant. At that time it was decided that Lockerbie Academy, which had not been hit and was accessible, should be used to provide temporary office accommodation for the various agencies involved in the incident. 'It was fortunate that the school children had just started their holidays so the school was vacant. This building had enough space for all the agencies to operate under one roof, and also had canteen and kitchen facilities.'

A first priority for police operations was the recovery and identification of the bodies of the victims; and two hundred and forty were recovered by the police, the army and the search and rescue dog teams by Christmas Eve. Although the cause of the disaster was not yet known, a criminal act was suspected which would require a massive police criminal investigation. The immediate implications were that pieces of the plane, cargo and the personal belongings of the passengers had to be collected as evidence by extensive line searches across the whole area. At that time it was not known what evidence would be significant to the inquiry, so practically everything that was found was collected and recorded.

Search and recovery

The search operation was obviously of such a magnitude that it would require far more personnel than was available in the small Dumfries and Galloway Force (at the time under 350 officers). Assistance was requested from other police forces and from the armed services and in the few days immediately following the disaster approximately 2,000 police, emergency, military and volunteer personnel were working in the area.

The search for bodies and wreckage, of course, started immediately. A primary objective for the police at the scene of any crime or accident, is to delimit the 'locus' so evidence can be preserved. 'You place a cordon round the area and "sanitise" it so nobody gets in. You try throwing a cordon round 845 square miles in countryside like that! It's impossible! Attempts were made to cordon off in the normal way but many aspects of normal police practice had to be compromised ... really it was unlike any "normal" disaster because the area to be controlled was just too big. It was impossible to keep people out.'

Debris from the plane was spread from coast to coast and over the north of England and a body was located six miles distant from Lockerbie town centre. The huge area to be searched was divided into six named 'sectors'. The operation of the search and recovery in each of these sectors was managed by a Detective Chief Inspector with a team of detectives and uniformed officers. Each sector team had police and army personnel for line search purposes as well as access to other specialised police officers and technicians (the mortuary team, video and stills photographers, scenes-of-crime officer, forensic scientists, officers of the Air Accident Investigation Branch, and medical staff). 'This description may make it sound like a table-top exercise. Suffice it to say that in reality it bore little resemblance to that. For a start, because it was a murder inquiry we didn't know what debris would be significant. As a result virtually everything had to be maintained, labelled and recorded.' The searchers did not know what they would find. 'Although they were told what to expect, as they set off, they didn't really know they would be finding children under five, or old people of about eighty. They didn't appreciate that they were going to find human remains of children, of old people, of souls entangled in aircraft wreckage. And people impacted in the ground. This is what the cops were coming up against. Also, the hilly terrain with marshlands, little lochs and forest around Lockerbie was very difficult to search. There are very few roads, and those that did exist were narrow, winding and built only for light use. Because of the time of year, daylight hours were only from about eight in the morning to three in the afternoon and so the working day for searching was limited. Quite probably more pieces of the aircraft and personal belongings of the passengers will continue to be found for many years to come.'

Police control of the influx of people into the area

'A huge problem in the immediate aftermath is that people cannot readily make contact, and as a result they will just turn up in droves'. It was established within a very few hours that there were no survivors from the plane. The emergency response was obviously in anticipation of a far greater number of casualties. As it was, eleven residents of Lockerbie were killed, two residents were seriously injured and three others were treated for shock and minor injury.

'But there were something like thirty ambulances in Lockerbie. Although the fires were brought under control very easily, there were five tenders in the area. In many ways it is quite difficult to stop the emergency machinery once it is in place, especially if communication is blocked. Plans co-ordinated by Regional Emergency Planning are already in place to respond to a civil emergency of this magnitude, and all the different agencies really have their own agenda.' It is only through the police taking control of the situation that necessary emergency and welfare assistance can get access to the area. 'From my experience a fireman arriving at a fire will not neatly park his tender, but will stop it in the middle of the street and may not give adequate consideration to disrupting other traffic. His priority is to fight fire. On the night of the disaster a fire tender broke down because the tyres were burst by debris, and some ambulances ran out of petrol.' Because of this the police need to control not only the enthusiastic volunteers, but also the activities of other emergency workers and their vehicles.

'People continued to arrive from all over the place. In the course of those first few days I learned about all sorts of organisations I didn't know existed. For one thing they all talk in abbreviations and acronyms – it's all "telecons" and that kind of stuff and you spend half the time trying to figure out what the letters mean. It is essential to liaise with whoever is in charge of these groups in order to manage and organise their activities. There are great difficulties in managing all the people who arrive fired up with enthusiasm and wanting to help. They all come with a different agenda, and they don't necessarily appreciate what the basic requirements are or what priorities have been set.

Understandably, all they want to do is get out and *do something*, their adrenaline is pumping. Many of them are volunteers, they are there to do what they can, they're anxious to assist. You have to be brutal even when it is obvious that they have made a great effort to get there. It's either, "Yes we need you, or no we don't". If you don't grab their energy and channel it then there is absolutely no doubt about it, they will go and do their own thing. And that is what happened in too many instances.'

The police must also communicate with the voluntary agencies – many of whom turned out to be essential to the operation. For example, communications could not have been opened up without the Raynet amateur 'radio hams' system. 'We were able to communicate through these local radio enthusiasts and while their efforts are to be commended, someone had to be stationed along with them to make sure the right message got sent. You can appreciate that the message can get quite distorted being passed along the line.' It was also impracticable to carry out security screening of these volunteers, and this was a bit of a concern for control and security purposes. But the police and armed services acknowledge the help of the voluntary organisations. They believe that the search and recovery was carried out much more quickly, considering the demanding terrain, than would have been possible without the search and rescue dog association (SARDA). 'The Women's Royal Voluntary Service (WRVS), and the regular kitchen staff from Lockerbie Academy worked a three-shift system to provide meals to the more than two thousand personnel working at the site. The Salvation Army provided soup, coffee and whatever else from their conventional soup kitchens to meet up with searchers and the diving teams searching the lochs. The search personnel who had trekked for what seemed like hours over the hills were quite amazed and cheered to walk over a hill or emerge from a wood to find the Salvation Army there. Somehow they managed to get their vans into quite remote areas.'

Media

Within three to four hours of the disaster more than a hundred media representatives had arrived at Lockerbie and at its peak this figure rose to six hundred. 'By the early hours on 22nd of December 1988 Lockerbie was already awash with reporters and camera crews all desperate to meet their editors' demands for information. And never underestimate the speed and technology available to the media. The American station CBS picked up the Lockerbie story over the wire. They immediately chartered a plane at Heathrow, phoned Carlisle Airport and arranged for vans and cars to meet them there. Within two hours of the impact, they were ensconced in Lockerbie Academy beaming pictures live to the US by satellite.'

It was essential that the police exerted some control over the activities of the press so they did not hamper the progress of the investigation. 'Particularly that night it was very hard to know what to do with the journalists until something could be organised, to keep them out from underfoot. Given the intrinsic rivalry between the competing newspapers and broadcast stations some journalists went off and did what they wanted. The media had had obvious difficulties in making direct contact with a spokesperson from Dumfries and Galloway police who – at the time – had no dedicated press office. The normal call-out procedure for staff of the Strathclyde Police Force Information Office was followed and a press office was set up in Glasgow in liaison with Dumfries and Galloway. Given their long established links with the Strathclyde Police press office, however, the Scottish media were already making contact with them.'

'Early the next morning the Strathclyde press team was transported to Lockerbie Police Office where they established a World Press Centre. Later, a public hall adjacent to Lockerbie Police Office was taken over for press conferences and a daily programme of briefings was organised. Media people have a job to do and that in large part means filling a limited time spot in a news broadcast. Once this requirement is understood by police management, appropriate material for the press can be organised thereby diverting them from doing things which potentially could hamper the investigation. With adequate provision of this material, the media can fulfil their function and disseminate information to the public.'

'Having regular predictable press briefings had several advantages. For the police it kept the media away from the Incident Control Centre since the press office was some distance from it. From the journalists' perspective it made the task of gathering news easier. The system worked so well that eventually it was accepted by all Local and Central Government agencies that the press conference centre should be utilised for all media releases. This allowed a consensus between the police and the government as regards the nature and timing of any release of information. Because the press knew the arrangement, as the days went on many camera crews simply left their cameras in place in the conference room between programmed sessions. Whenever there was a need for location shots the police team organised "pool" facilities whereby a set number of reporters, stills photographers and cameramen were taken by police transport to specific locations, such as the cockpit area at Tundergarth or to photograph the line searches to obtain material to be shared by everyone.'

'The regular briefings kept them fairly quiet, and were largely very successful, but some were still out there being quite intrusive. The search teams still found some photographers popping up from behind hedges with long range cameras to film rescue personnel struggling with bodies. It was very upsetting for the police officers to have the press poking cameras at their back, but as it turned out the abundance of material provided by the police press office meant that very little of this "maverick" footage was actually transmitted.'

The relatives of the deceased

'Special dispensation was given to recover the bodies as early as possible. One of the reasons was that the bodies were lying all over the place and the decision was made to move them rather than leave them at the locus for forensic and criminal investigation. The relatives of the deceased who could get there appeared almost overnight. As you would expect, all the relatives wanted information about the disaster, about how their loved one had died and what was being done to find out what had caused the catastrophe. They also wanted access to the body. It was important that the police balance humanitarian concerns with the practical and legal issues of body identification and release.'

The Chief Constable decided on a policy of personal contact with the relatives and a unit was set up for this purpose. This personal contact and having an individual named officer dedicated to a particular victim's family made it easier for the relatives and for the police, although one officer working in this unit said, 'In interviewing the relatives, I will never forget the sadness, shock, horror and disbelief on the faces of the American relatives and of the people of Lockerbie.' There were people of twenty-one different nationalities on the plane. Very early on the police had to familiarise themselves with different cultural approaches to bereavement. As an example, 'There were some Japanese relatives who had built a shrine on one of the road-sides and were worshipping by it. We knew nothing about it until a police patrol found them. Obviously this situation would have to be controlled but with great sensitivity. This is just another example of the range of unexpected demands placed on our personnel at the site.'

Decisions which at the time appeared humanitarian led to great difficulties for the individual police officers assisting the relatives. 'The relatives were not allowed to see the bodies (because of their condition), so many of them wanted to see the actual place where the body of their loved one had fallen to earth. Bodies had hit the ground at a hundred and twenty-eight miles an hour which in some areas had caused body-shaped craters in the ground. The police decided the best thing was to get the craters filled in and back to normal as quickly as possible. So, as soon as the areas had been cleared of the victims and the property and the aircraft wreckage the District Council came and smoothed over.'

'What then happened was that the relatives arrived and wanted to see exactly the spot where their loved one had been found. Several relatives wanted to see the crater and touch it. It was unbelievable. It's something we would never have expected. It seemed that they had nothing else tangible to help them out in their grief. Obviously it was a real problem for the officers taking the relative to the spot if it had already been filled in and was no longer visible. It put the officers under great pressure.'

'It is understandable that all of the relatives wanted information. But because of the criminal investigations and the extent of the inquiry, information had to be controlled. The Deputy Chief Constable of Dumfries and Galloway made it a point to meet relatives two or three times a day to try to meet this demand. This policy also relieved the areas of the operation which were under "heavy pressure" from relatives. For example, relatives wanted to talk to the line search teams in the fields and the officers working in the body identification area were also under pressure from the relatives.'

The mortuary

The police have a very specific role in documenting any sudden death in the community. At Lockerbie a team under the direction of a Detective Chief Inspector was set up to process identification, complete sudden death reports, and register the death before release of the bodies in coffins to relatives. The police, along with the Army and specialist search and rescue volunteers, also played a major part in locating the bodies in the town and on the hillsides. A helicopter was used to transport bodies, in body bags, which had been located at a distance from the town.

'Initially an emergency mortuary was set up at the town hall. On the night of the disaster, the occasional person had found a body or a body part and they obviously needed to know what to do next. So what happened was that they came to the town hall, at the time it seemed like a natural place, being the centre of this small town for the local people. The decision to use the town hall for the mortuary was almost made for us and it was a decision over which we had little control.'

'As it turned out, while the town hall was able to accommodate eighty bodies, as more were brought in space proved a challenge. A further difficulty was the physical lay-out of the town hall, which was virtually on three levels. Post-mortems were conducted in the lower part of the building and the biggest physical problem was carrying the bodies on stretchers down the winding staircase into the basement area and back up again afterwards. Some of these bodies were very heavy and, because of the way in which they had died, were physically unstable. What should have been done is always clearer in retrospect, but a less suitable place for a mortuary than the town hall would have been hard to find.' In some accounts of the disaster it is described as 'luck' that the town had an ice rink. It is true that many towns do not have one, and in the circumstances the ice rink made an ideal second mortuary. In addition to these two mortuaries further empty factory space was used for the storage of coffins containing bodies prior to release to relatives. The coffins needed to be lead-lined because of regulations about the return of bodies to other countries. This obviously made them extremely heavy and hard to handle. 'There is no doubt that working in the mortuary was the most demanding of the police officers' tasks at the disaster site. In the mortuaries they carried out the whole range of horrendous tasks including stripping the bodies when they arrived, and assisting (by taking notes) during the pathologists' examination.'

Personal property

In order to ensure there were no further bombs amongst the material from the plane, a mobile X-ray unit was used to examine all baggage and cargo before any item was opened. The baggage was then searched to identify it with a particular passenger, all property was labelled and stored and any valuables deposited in safes. A large quantity of items of personal property was recovered and taken to a local warehouse for safe-keeping and examination for evidence. Officers in the property department continued to work there well into 1990, returning property to relatives. The officers found handling the personal property, and especially the clothing (all of which had been washed, repaired and ironed by women in the community) of the deceased very demanding. Their job entailed a great deal of contact with the bereaved relatives and even in this task coming face to face with the reality of the lives of those who had been killed.

Some recommendations resulting from the Lockerbie operation

People involved in disaster management and recovery say they wish they had known what to expect and all disaster workers feel this uncertainty. But every disaster is unique, and at Lockerbie the only realistic approach was to expect the unexpected since it is impossible to anticipate everything. There are, however, issues which confront most organisations in the aftermath of disasters, and the following is a brief summary of them.

1. The dissemination of information

A disaster throws all normal channels of communication into chaos and, coincident with that, everyone who is involved is desperate for information. Families, residents, the media and all those working at the site need to know what is happening, and to be reassured that progress is being made. The police are required to facilitate and organise this.

Staff also need information. 'Because of the magnitude of the operation and with so many officers all over the place, this information was not as easily available as it should have been. Many of them felt cut off and deprived of information. We learned that even though it is time consuming it is essential to morale and control. So making sure they are well briefed is crucial and should be done as frequently as is practical. They also need to be informed when they are temporarily stood down, at ease, until they are needed. Rumour which tends to develop in an information vacuum needs to be managed and controlled. One solution is to provide a regular news-sheet to at least inform the gossip, and stop the scaremongering.'

2. Protection against health hazards

It is important, especially because of the sort of demanding tasks that the officers have to do, that they are looked after. This is important both to protect the workers from health hazards but also to keep morale as high as possible. Access to medical assistance and first aid is essential. Physically there are dangers from sprains and strains when lifting bodies and other heavy material unassisted.

And working on rough terrain presents potential problems of stumbles, strains and fractures. When there are large numbers of officers working with bodies, body parts and potentially infected material, there is a pressing need for careful personal hygiene. Toilet and washing facilities with hot water and soap are essential to control infection.

Workers must also be adequately protected against cold and rain, as well as having the appropriate footwear for the job. Officers were issued with Wellington boots which were appropriate for most purposes but proved inadequate in others. 'There are also lots of uncontrolled aspects to a major and complex operation like this, and additional problems and hazards come in unpredictable ways. There are all sorts of things, wee daft things that you never think about or expect, that are going to upset the workers. For example, part of the cargo on the Lockerbie plane was a big cardboard box of hypodermic needles. Obviously we didn't know beforehand that the cargo contained a box of needles, and we didn't know what they were. Seventeen officers had their feet punctured because of standing on needles and naturally they were worried about whether they were used, contaminated needles. We got them suitable footwear so that the needles couldn't get through the soles of their Wellingtons, and they all had to go and have an anti-tetanus jab.'

3. Provision of food, rest and appropriate transportation

Arrangements need to be made for periods of regular rest in surroundings that are as warm and comfortable as possible. The workers also need hot food, which even although there were so many workers at the site was not a problem at Lockerbie because of the volunteer help. The disaster site was over sixty miles from either Glasgow or Edinburgh, the main centres of population, which presented problems getting the large number of personnel to and from the area. 'Officers left their homes at perhaps 4.00 a.m., travelled for two hours to Lockerbie, worked on the hill all day out in horrendous conditions, and returned cold, wet, tired, dirty, and distressed. But the number of times coaches broke down, or the heaters didn't work – particularly bearing in mind the time of year. It wasn't good. I lost count of when we had to go and hunt for a bus because it was late or missing. You need to spend the money, you can't cut corners on these things.'

4. Control of the huge numbers of people wanting to help

The emergency services as well as volunteer help need to be checked, managed and controlled. The point has been made that, without this control, the rationalisation of what is required and its provision would be hampered. Finding out what a potential helper's particular expertise might be and who the leaders are is essential to the decision of whether they can be of use to the overall operation. Often the emergency service workers feel helpless and impatient during the inevitable times when they are waiting for orders. 'There is nothing more demoralising for workers who want to be active and not getting a chance to do anything, and being told to just sit there.' Again informing them of what is required, and making it clear when and where they are required is essential at regular briefings and through the appropriate channels.

5. Care for people in distressing circumstances

A disaster inevitably produces a large number of people who are, to one degree or another, traumatised by the event and they need to be treated sensitively and supplied with information and social support. While the police obviously do not provide this support themselves in any 'official' way, they do need to facilitate the work of social service agencies and bring people whom they come across in need of help in contact with those who can provide it. In addition, under chaotic conditions, a person in a uniform and in particular a police uniform is a comfort as well as the focus for those seeking information and help.

The police themselves are also in distress when working under these conditions, and the unusual degree to which they are required to handle the bodies as well as have contact with the relatives of the deceased places particular demands on them. Volumes are already written on the subject of stress in people working at the site of a disaster, but in practice the considerable pressure staff are under is not always evident even to the workers themselves. This phenomenon may be enhanced in emergency workers, including the police, whose *modus operandi* is to be tough and resilient. While the job is being done, the spirit of teamwork and the engagement in the work may mask symptoms of stress. Workers may have difficulty sleeping, or may drink alcohol to help sleep leading to cumulative problems. They may feel so committed to the task that family and other considerations take second place, and normal social relationships and sources of social support are disrupted. The emergency worker may feel that such noticeable changes in behaviour are 'normal' and attributable to the nature of the task. They may not wish to take a break because of being highly committed to the task and a desire to see the job through to completion. This level of activity is also a well known protective strategy in the face of distressing tasks such as in the mortuary or on line searches.

This was evident at Lockerbie. It was assumed good management practice to rotate officers performing unpleasant tasks, but it became evident that this was not the simple solution it appeared. While the number of hours of uninterrupted work, particularly demanding work, needs to be controlled, distress felt by staff is not always minimised by taking them off difficult jobs. Indeed rotation out of tasks may be associated with its own stress. 'The officer's way of coping was to see themselves as part of a team, geared up to get the job done. Along comes the caring management wanting to rotate them. But they were upset because they saw it as their own job being taken away. They wanted to continue, to see it through and sort it out. We had misunderstood and we realised that rotation is not always the answer. Having learned that, we thought it would be better to ask for volunteers to leave the mortuary if they had had enough ... but again they were not going to say so in front of colleagues. So we gave them the opportunity to do it confidentially. Not one volunteered – every one of them stayed.'

'You have to be aware of the pressure these guys are under. In briefings and in your own talks with their supervisors you need to make them understand that you are not spying. We tried to provide some welfare by passing it around that there was a room where anybody could come along for a confidential chat, a cup of coffee and a cigarette. But no one used that or 'outside' help from social work either. But I know from informal chats with some of them that there were some who had a lot of trouble, who were really affected. It is very hard to know how to provide support in the police, and we are still learning.'

Figure 2.1 Agencies in the Incident Control Centre at Lockerbie

---

Police
Chief Constable
CID
CRISIS
HOLMES
Senior Investigation Officer
Police Liaison
Operations
Traffic Control
Processing Bureau
Victim Identification/Registration

Special Branch
Federal Bureau of Investigation (FBI)
Maps
Procurator Fiscal
Dumfries and Galloway Regional Council
Air Accident Investigation Branch
Raynet Communications
Search and Rescue Dog Association (SARDA)
American Embassy/Consulate
Army Liaison
Social Services Agencies
British Telecom

Figure 2.2 Police Forces providing mutual aid to Dumfries and Galloway Constabulary

Strathclyde
Lothian and Borders
Central Scotland
Grampian
Metropolitan
Cumbria
Northumbria
West Yorkshire
British Transport Police
Ministry of Defence Police

Notes

1 Thanks are especially due to Ex-Chief Superintendent Ian Dickie, Dr W. D. S. McLay, Chief Medical Officer, and Constable Ralph Howden, all of Strathclyde Police, Superintendent Carpenter of Dumfries and Galloway Constabulary, and many other Strathclyde officers of all ranks who took part in the recovery operation at the Lockerbie site.

3 Social Services after Hillsborough: managing the initial response

David Mason

David Mason is currently Director of Social Services for Warwickshire. At the time of the Hillsborough Disaster he was Director of Social Services for Liverpool, and was the Chair of the Inter-Agency Group that was set up to co-ordinate the welfare response to the tragedy. In this chapter he describes his role in the immediate aftermath of the disaster and focuses in particular on the demands of the media and his position as the central spokesperson for the conglomerate of social services departments that responded to the disaster.

The day and the initial response

Driving home, listening to the car radio the football commentator seemed for some incomprehensible reason to be playing a recording of the Heysel Stadium Disaster instead of describing the two FA Cup semi-finals in Birmingham and Sheffield. After a few moments I realised that the report was real, that these horrendous events were actually happening and being beamed directly into homes and cars. The seriousness and severity of the carnage was in even sharper relief on television and for a short time after arriving home I watched and listened increasingly aware of the need to respond.

A telephone call from an off-duty manager from the Emergency Team required the first decision to be made; the team on duty were to co-ordinate information and action, while she investigated what links could be established with Liverpool Football Club. On the Saturday afternoon of the semi-final the answer came back very quickly – none: everyone was at Hillsborough for the match. For the next four hours I was unable to leave the telephone. Merseyside's Chief Probation Officer was the first caller, followed by numerous colleagues from Liverpool, and other social services departments. The message was the same from everyone – we want to help, what shall we do?

Establishing links with other people in both Liverpool and Sheffield became more and more difficult as the day wore on. Telephone lines were either blocked or numbers engaged. Everything seemed to take an eternity to complete and there never seemed to be enough information to answer any question. Even a friend of one of my sons arrived at the front door distressed and seeking information as he had obtained tickets for people who were in the crush. After talking to my opposite number in Sheffield, my deputy, the Chair of Committee and Leader of the City Council and many members of staff, a team of social workers and the Council Leader and myself set off for Sheffield. The social workers were all members of the same team and were experienced child bereavement counsellors. With their senior they had volunteered to establish an operational link near the disaster. They, more than I, anticipated the value of having a clear picture of what was happening in Sheffield for future work with families and survivors. The Leader in accompanying me was keen to have some basic and first hand knowledge for planning and organising our response. It is difficult to convey how little we knew and how quickly the news was changing and always for the worse.

Our journey was uneventful and passed almost in silence except for the few calls to the Leader's mobile phone. Neither of us seemed to know what to say or precisely how to approach the task of establishing the dimensions of a disaster. We started at the office of the emergency duty team in Sheffield, and trawled what data they had. They were receiving reports of casualties and the grim reminders of the seriousness of the disaster in very personal terms. I wrote the details down and much later that night saw a family searching for their son, not at the hospital, but at the gymnasium which was in use as a temporary mortuary.

It was agreed that we should move on to the youth club which had been established as a receiving centre for families. The awful reality of the day took on a new poignancy. The club was full of people from Merseyside and further afield seeking news of their sons, daughters, husbands, wives, fathers and mothers. In almost equal numbers were social workers, clergymen and voluntary workers; all there trying to help. The team of workers from Liverpool had arrived, like us via the emergency team HQ. They were now engaging in two tasks – collecting information and making contact with Liverpool people. My memory of this time is one of intrusion – what right had I to be there – my family were safe.

We went on quite soon from the youth club receiving area to Hillsborough, to the temporary mortuary at the gymnasium. Our arrival there was without doubt the biggest shock either of us experienced in a day which was characterised by shocks. This sports centre was full of people both living and dead, never before or since have I felt so much pain and grief. I had my first understanding that a disaster with almost a hundred deaths is somehow different to one hundred separate deaths. This realisation was reinforced many times over subsequent weeks and months.

As well as the team from Liverpool there was another social worker from the city in the gymnasium/mortuary, who had been amongst the first to answer the radio appeal. The staff on duty were working in pairs, social worker and clergy, supporting families here to search and identify their relatives who had not returned from the match. In this vast space about half the floor area was covered by the rows of dead bodies. It seemed to be overflowing with silent and numbed people. We studied the growing list of the dead who had been identified and the staff group talked to local social workers about how families were being helped.

Our next brief visit was to the church hall which was in use as a centre for those families who had unfortunately found their relatives on that gymnasium floor. While we were here one of the people from Merseyside recognised the Council Leader and exchanged a few words with him. There was little we could do, having I think grasped the nature and to some extent the enormity of what had happened that afternoon at Hillsborough.

If conversation was limited during the outward journey, it was non-existent during the return. Neither of us said a word for about an hour; each trying to come to terms with what had been seen and heard. Eventually on the outskirts of Manchester the question I most dreaded was asked – 'What should we do?' I had been trying to remember everything, anything I had read or heard about responding to disasters, while at the same time my thoughts continually returned to the Hillsborough gymnasium. My reply included provision of a helpline to ensure easy access to people seeking help; a consistent social work service to help individuals and families cope with the consequences of disaster; a drop-in facility for open access, and co-ordination of the service to ensure all departments were in accord. Fortunately we got lost finding our way around some road works and I was not pressed particularly hard. I was reassured to be told that I would not have to go through the usual Council committee rounds for approval prior to action. I had mentioned this to the Leader as a key component identified after the Bradford Fire.

I left the Council Leader at the studio of a local radio station in Liverpool and then drove the last twenty miles home to a sleeping house, getting in just after 4.30 am. Physically tired but with my mind racing, I sat and tried to make sense of what had happened and what I might do to organise a service in the wake of the disaster. Without any warning I was suddenly sobbing, thinking of my son almost 14 and the details I had written down at the Sheffield emergency duty team office about a boy of about the same age. A brief sleep, a shower and then a telephone conversation with my deputy, before leaving at around eight o'clock. My deputy had arranged a meeting of social services departments in Merseyside, together with representatives from the health and probation services, the Emergency Planning Officer and Social Services Inspectorate, as well as advice from 'Crisis Psychology' – specialist disaster consultants.

The office was already open when I arrived, but the porter was not in evidence. Two reporters from a national newspaper had beaten me to my room, though they left after I had made a commitment to hold a press briefing later in the morning. I had not experienced this level of interest in the work of the social services department before, and I had a lot to learn about the way the media behave and react to a major disaster. At times during the next week I was tempted to conclude that the work of the department was secondary to the need to report it in the press and on TV. By ten o'clock everyone who had been invited was assembled, together with scores of staff who had turned in to work and wanted to begin to offer help. A group of senior support staff were available to deal with the logistics and it was time to get on with the job. Just before we began I had a few minutes with the partners from Crisis Psychology. One of them, Michael Stewart, explained what they could offer, how he had worked in Bradford, and some of possible difficulties. He warned that one element of a co-operative response to a disaster was competition between the agencies during the early days, followed by a desire to withdraw when the initial excitement was past.

The first meeting went well, everyone appeared committed to working together and delivering a consistent service that transcended Council boundaries. At this stage we had not grasped that the dead, the injured and the traumatised people from Hillsborough were drawn from a much wider area than Merseyside. It was agreed that we would provide services jointly:

- helpline – as primary point of entry and assistance;
- database – to record both users and offers of help;

- outreach service – near Anfield to provide open access;
- co-ordination – through regular liaison between agencies;
- staff training – arranged jointly with Crisis Psychology;
- public relations – one spokesperson for the response.

Given the centrality of Liverpool it was agreed that I would act as both co-ordinator of the social services response and media link.

At the end of the meeting the corridor and indeed the whole building was full of people. The computer experts had diverted telephone lines to provide a bank of phones for the helpline, temporarily arranged in the general office. The City Treasurer was advising me of the amount of cash that was instantly available, should it be required – an unusual occurrence in itself. The Leader and other Members of the City Council were waiting to be updated on plans and progress. The media were collecting at the entrance to the office for the promised mid-day briefing.

Having prepared a short statement, which I discussed with Michael Stewart, I descended the stairs and gave the first of many interviews. The prepared text had to be repeated many times as most television crews wanted the briefing to appear like a discrete interview with their station. The last interview was with a group from Belgium, and their final question really threw me – 'Did I think that Hillsborough had happened in some way as a consequence of Heysel?'

This process had eaten into the afternoon and time was running away. The first of many briefings and training sessions for staff was already under way. Liverpool senior staff team had not found time to organise and plan our own work. We quickly agreed responsibility for each activity, a decision we frequently revisited as the operational pressures changed. At this point no-one was giving much thought to the 'normal' work of the Department which we had to maintain while this was going on. I made a telephone call home requesting suitable clothes for the service in the Metropolitan Cathedral, as a result of councillors deciding that I should attend; this interrupted planning future work that afternoon. The helpline was already being used to both request help and offer assistance. Numerous calls were coming through on other extensions in the building, making co-ordination more difficult.

Our inter-agency review meeting later that afternoon included a representative from Cheshire, increasing to six the number of authorities now involved. We reviewed progress recognising the need to bring to bring together politicians from the collaborating local authorities, who needed to own the agreed response to the disaster. This meeting was being fixed for the following afternoon with the Deputy Leader of Liverpool making the arrangements. Overall it felt as if we had made a reasonable start, and we had reached some agreement about our values: these included being inclusive but not intrusive, offering help only if we could sustain it, reaching out to people, actively offering our help and being practical, not concentrating on counselling. We recognised that our counselling service would not be used until later.

After a change of clothes, Pauline Farrell the Director of Sefton Social Services and I left for the service, but failed to realise how long this short journey would take. Her intimate knowledge of the cathedral was more than useful; we managed to park her car in the underground car park and arrived inside the cathedral and inside the crowd. We were, however, late and made our way to seats near the front for civic guests as the Archbishop was beginning the service. I could not help wondering how so many people had come together so soon after the disaster happened and with so little publicity. As I left the service I saw a hand-made sign advertising an independent counselling service, which made me question if we could really avoid convergence and multiple visiting of people by different agencies – competition?

When we returned to the office I checked the organisation for the rest of the day, overnight and for early Monday. All day it had been difficult to obtain information from South Yorkshire police about the dead or their next of kin. Our efforts, and those of the Merseyside Police failed to elicit the data which was crucial if we were to deliver a co-ordinated response. The helpline, as well as receiving offers of assistance had also received fifteen direct requests for help. There were enough volunteers to staff it overnight and the next day. The logistics needed further attention and the Helpline had to be relocated away from the departmental general office. It was, however, already beginning to prove that it was an effective way of providing access to people.

As I drove home, alone in the car, I was aware for the first time of my tiredness and anxiety. The plans for the following day included a meeting of City Council chief officers with the Leader, deputy and other senior politicians; a press briefing; a further meeting with managers from other agencies and what else . . .? I did not know what to expect, but at the same time was being asked to lead and manage a disaster response. Through the day I had had snatched conversations with Michael Stewart from Crisis Psychology which had not reassured me. He warned me about the competitive behaviour of both individuals and agencies which would quite soon be replaced by disengagement and withdrawal once the media interest reduced. The responsibility seemed awesome and I doubted my own ability to do the job then, and many times during succeeding weeks.

The first week proper working on the response to Hillsborough began just before six o'clock on Monday morning. I was retracing my journey, having spent a few hours at home – some sleep, after some time with my family, but everything interrupted by the telephone. When I arrived, the office was already a hive of activity. Some staff had been working through the night sorting and planning as well as the staff operating the helpline, which had been taking calls. A review of the previous day's events, which had occurred overnight, was reported and then preparation for the day at hand began with a management team meeting at 7.30 a.m. This early start was to be the pattern for the whole of the first week, with the exception of the driving in to the office; the Council provided taxis because of the long days.

It became apparent immediately from the lists of the dead and injured that the boundaries of the disaster were not local to Liverpool, Merseyside or the North West, and contacts would need to be established with other departments throughout the country via the Association of Directors of Social Services (ADSS) and the Social Services Inspectorate. The people who had died and were injured came from all around the country although we had no precise information about either group, and it appeared that the media were given better information for their columns than we were for our service.

On Monday we were still at the stage when everyone wanted to help and discuss what they had to offer. The chief officers' meeting was quite straightforward – within Liverpool the social services was to be the lead department in respect of Hillsborough (to be put to a special meeting on the following Wednesday); a sub-committee was to be formed and arrangements made to organise a Disaster Fund. At this time the Council was without a Chief Executive, and so being the lead department meant we operated without any corporate framework.

Later that morning I met with the City Solicitor, the City Treasurer and the former Chief Executive to discuss the formation of the Disaster Fund. The ex-Chief Executive had been asked to administer the Fund. In discussion I was clear the social services role was to advocate for individuals not to become part of the administration of the Fund. Over the next few days the form of the Fund was agreed and it was helpful that the Merseyside authorities were not included amongst the Trustees, as there were already signs of conflict between members about which councillors and Councils should be included; competition was not confined to councillors; workers and managers also became involved in their own disputes.

The media

For the first week the media consumed a great deal of time; being a spokesperson in the aftermath of a major public disaster becomes a full-time job in itself. The City Council responded by attaching a full-time press officer and minimising the demands on me for anything but press conferences, briefings and interviews. Even this issue was the subject of some adverse criticism as it was assumed by some individuals that Liverpool, rather than all the agencies providing assistance and resources, would receive the credit and acclaim. I tried to be very careful and substituted Merseyside or North West for Liverpool; referred to Liverpool FC as 'the club', and mentioned every other local authority by name, no matter how small their contribution, in order to avoid further comment and criticism.

After Monday, when the press conference was held at mid-day, the daily session with the media became my nine o'clock appointment. On several days this was the prelude to a whole morning of interviews. The relatively early time enabled the local evening paper to meet their deadlines and I was conscious of the need to have a continuing rapport with the journalists of the *Liverpool Post* and *Echo*, and to ensure the arrangements were satisfactory for them.

For the first time in my career I was asked to do live television interviews. Although initially daunting, I soon realised they offered a unique opportunity to get points across. In recorded interviews very little material was actually used – about five seconds for every minute recorded. During one live interview I was able to get the Helpline telephone numbers on screen in spite of the question not being asked and the interview having finished.

I have already mentioned the difficulties of gleaning information from South Yorkshire Police. At six one morning when I called into the Helpline on my way into the office I found a worker in some distress. It transpired that she had spent a long time on the telephone talking to a woman whose husband had been killed. The woman had asked for her husband's wedding ring, only to be told by the police that it could not be returned for forensic reasons. This seemed an irrational and bureaucratic response to this grieving widow, which defied sensible explanation. As the contact with this police force was so unproductive, I decided to challenge the police to justify their behaviour during the press conference. This, as I hoped, brought forth not only interest from the media, but more importantly but perhaps by coincidence a positive response from the police, who agreed to release personal effects.

It had been agreed between all of the agencies that we would only have one spokesperson, but it had not been agreed how we would select what to share with the media. It was further complicated by a comment made by some politicians about the services being provided and responsibility for them. It felt as if I was walking on egg shells and I became anxious about how I handled the media, and what both I and others said. My handling of the matter of the wedding ring was approved of by colleagues as it fitted within our policy. We had decided to maintain confidentiality except where we had permission for disclosure, and to use the press briefings to air issues which individuals and families had experienced, and raised with social services staff. Interviews for both TV and radio which also involved survivors, I found especially stressful, as the environment was artificial, exposing the distress of traumatised people in public and then apparently discarding them to go on to the next item in the programme.

During these early days certain newspapers were making claims, seemingly supported by South Yorkshire Police, about the behaviour of Liverpool supporters at Hillsborough. This caused great distress to both survivors and families of people who were injured and who died. The reports were wildly different to the accounts which social workers heard in counselling sessions and over the Helpline. The dissonance between these reports and the experiences of people at the match created confusion and in some instances despair. The desire of some journalists to intrude into the grief of families seemed boundless, and the lengths to which they would go appeared to fall well short of reasonable behaviour. There were regular reports of people posing as social services staff in order to gain access to both homes and confidences. It was assumed these people were reporters.

By the end of the first week all staff working in all agencies on the Hillsborough services had to be issued with a new and specific identity card. On the Friday a response and rebuttal to the press reports about the behaviour of the Liverpool fans was the subject of an extra press conference held late that morning. The Taylor Inquiry subsequently refuted the claims about the behaviour at the ground. To my amazement, I was contacted by reporters from the local newspaper who wanted to both distance themselves from the excesses of some of their London-based colleagues, and share their experiences with me. One person actually gave me an on-the-record quote about the behaviour of journalists in Sheffield on the evening of the disaster. The approach of the Liverpool papers was faultless, they asked once for an interview and accepted a refusal if that was the result. During the press conference I felt I had considerable support for my views from most of the reporters present. One, who pressed me for evidence, was shouted down by the rest and left before the end of the briefing.

I have had a lot to say about the media during the first week after the disaster happened – I have asked myself why this is so? My memories of that week, which seemed to last forever, are dominated by the need to feed the demands of the journalists. After a disaster the extent of contact with the press should never be underestimated, as I think we did. I believe that the real work suffered as my time was diverted into this activity.

Other areas of work

What of the other real work? The helpline was running in its new location, with teams of staff from a variety of agencies being co-ordinated by shift leaders, all middle managers from Liverpool. Their deputies were maintaining services in district offices and hospitals as well as taking a turn to work on the telephone. By Tuesday, it was clear that the single helpline approach was in danger of breaking down – one department had developed its own publicity material, which included a separate telephone number. This created pressure for other teams to give out their telephone numbers, and the clarity which we had originally struggled to achieve for the public only forty-eight hours earlier, was in danger of disintegrating. I was both surprised and worried by this development and spent a lot of time over the day and evening trying to have the offending material withdrawn. This was eventually agreed and the obviously uneasy consensus continued intact. This was important if the service was to be both credible and available to people from around the country who might need to use it.

One of the unusual features of the service which developed was the outreach work around the football club. We quickly established a drop-in centre near the Anfield ground, at the Vernon Sangster Sports Centre. This was supplemented by local drop-in facilities in some of the other authorities. However, no-one had foreseen the number of people who would wish to visit Anfield, and the outreach work was increased further by the deployment of workers on the pitch, and in the lounge and administrative buildings of Liverpool FC.

Thousands of people visited the ground and left flowers and donations to the Disaster Fund. Some of these visitors had been semi-final spectators, and some were actually in the Leppings Lane crush where so many had died. For many the nightmare was brought alive and workers were able to make contact with them and offer help. People who were injured, and the families of those who had died used the facilities of the club where social services and workers from other voluntary agencies cared for them. They were also close to the heroes of Liverpool FC, many of whom were frequent visitors, spending time with people and subsequently attending funerals.

The services we had set up were signposted from the end of the motorway, discussed by travellers on buses and featured in newspapers. The decision to try and deliver a full range of services with a high profile seemed justified. It also put pressure on everyone to ensure that we could provide what we claimed. We were determined to try and reach as many survivors as possible, as our thesis was that to deliver non-intrusive but inclusive services was a sound investment in preventive mental health. Seeking out potential users was for us an unusual approach – generally we rationed our services to the public, concentrating on families and individuals whose social situation was close to collapse.

Advice, which we received from Crisis Psychology, pointed us towards composing a simple leaflet which would be generally available to the public, making them aware of the availability of help and the reactions they might anticipate after a trauma of this magnitude. It was helpful to have copies of material produced after other disasters, and particularly, the Herald of Free Enterprise. In our enthusiasm to go to press with the leaflet we missed some errors in the text and omitted to acknowledge the use we had made of the Herald material.

Within days it was apparent that there were many children affected by Hillsborough and a suitably modified version of the leaflet was devised late one night by Liverpool's Deputy Director of Education, public relations staff and myself. The final item that was being prepared during the first week was a letterhead which had the logos of all the participating authorities for all of us to use, instead of letters being written on Liverpool paper. This was an indication of the support of all local authorities.

Everyone felt the need for training as the enormity of the task which confronted us unfolded. It had been agreed at the first meeting that we would all need to invest in the staff delivering the service, and Crisis Psychology were commissioned to deliver a programme which satisfied the needs of a range of workers. For the first few days the programme was focused on managers with the main thrust being delivered to groups of main grade workers over the succeeding weeks.

At the end of the first week two sessions were planned to brief local councillors. These were held in Liverpool Town Hall and Michael Stewart talked about the start we had made; the services we needed to develop and some of the problems that survivors might have. These were a great success in spite of the usual logistical problems of hired video equipment breaking down. One of the many emotional moments for me, during that emotionally charged week, was when a councillor who had been a spectator during a football stadium disaster several decades earlier spoke of the feelings he had repressed and only begun to resolve that morning. The afternoon session was stopped while the participants went onto Dale Street to observe the moments of silence for those who had been killed at Hillsborough.

As the developments began to take shape it was realised that a continuing service for people affected by the disaster would be necessary. We actually received many offers, including the building of a permanent centre close to Anfield. Within the inter-agency group there was also a realisation that there would need to be planned provision to take forward the initiative and sustain it over time. If the needs of survivors and families from Merseyside and around the country were to be met effectively then we could not rely on goodwill alone. There was a need to try to establish the scale of the service needs, specify how they might be addressed, and allocate the necessary resources. The inter-agency group agreed that a group of planning specialists should be brought together under an independent Chair to prepare a report by the end of the second week. A professor of social work with strong Merseyside connections agreed to undertake this task on our behalf.

In the latter part of the week my greatest single omission dawned – insufficient attention had been given to supporting staff who were undertaking the direct work. Some authorities had made individual arrangements; there was some debriefing for staff finishing their shifts on the Helpline but there was no systematic debriefing and support service available. Most staff had kept on working very long hours and dealing with anxieties and worries as best they could. A few staff were suffering and some had taken sick leave owing to the stress.

The British Association of Social Workers (BASW), through its Chair, offered to manage and co-ordinate this assistance for us, and the first group of counsellors were provided by his own authority. This was agreed when I spoke to his Director early on Saturday morning while the latter was still in bed. The team of staff, who all had experience of post-disaster counselling as well as training in the subject, arrived on the following Tuesday.[1]

Before going home on Saturday evening, Michael Stewart and I visited Anfield for the first time. there was still a queue of people waiting to get into the ground; many of them carrying flowers and scarves in the colours of their team. From the terraces I saw half the pitch covered with flowers, some wrapped in cellophane. The weather had changed and the only sound was the rain. People filed silently across the pitch leaving their flowers, scarves and donations. It was incongruous that a stadium so usually full of noise and excitement should be silent as a cathedral. This was my first appointment for the week outside Council buildings and TV/radio stations. It felt as if my life had been overtaken and taken over by the events of the previous Saturday – an impact seemingly shared by others at Anfield.

My first week of managing the service after the Hillsborough Disaster felt as if it had been punctuated with conflict, competition and disagreement. In spite of these difficulties we had begun to deliver a service; our collaborative work was holding together; there were many people who had a real appreciation of what we were trying to achieve; staff commitment was high and the press were beginning to lose interest in what we were doing, and this helped us to get on with our jobs. They had not, however, lost interest in raking over in public the pain and anguish felt by so many who were both directly and personally coping with loss and grief.

Notes

1 A personal account of the development of the initiative over the following two years can be found in Jane Harper's chapter in this volume.

4 The Kegworth/M1 air crash: setting up a helpline and a support service[1]

David Whitham and Tim Newburn

> **David Whitham** is a Social Services District Manager for Nottinghamshire County Council. At the time of the Kegworth/M1 and Hillsborough Disasters he was Group Principal Social Worker at University Hospital, Queens Medical Centre in Nottingham and managed the Nottinghamshire social services response to both disasters. He is the author (with Tim Newburn) of Coping with Tragedy: Managing the responses to two disasters (Nottinghamshire County Council, 1992).
>
> **Tim Newburn** is a researcher currently based at the Policy Studies Institute in London. He was previously based at the National Institute for Social Work where he managed a two year study of the social work response to the Hillsborough Disaster. This chapter reports David Whitham's experiences of managing the initial response in Nottinghamshire to the Kegworth/M1 aircrash, including putting together a team of social workers and administrators during the evening after the disaster. He concludes with a series of recommendations for social services managers who might find themselves in a similar position.

If you can keep your head when all about you
Are losing theirs and blaming it on you,
If you can trust yourself when all men doubt you
But make allowance for their doubting too;
If you can wait and not be tired by waiting,
Or being lied about, don't deal in lies,
Or being hated, don't give way to hating,
And yet don't look too good, nor talk too wise ...

Rudyard Kipling. *If*

The lead-up

I was watching a recording of the FA (Football Association) Cup match that had been played the previous day. It was when Coventry were knocked out by the non-League side, Sutton United. I'd recorded it because we'd been out on the Saturday night and I particularly wanted to see Coventry City beaten, as in one of the previous season's FA Cup semi-finals they had knocked out my team, Sheffield Wednesday. Anyway, this would be early Sunday evening. My daughter was upstairs watching *Top of the Pops* or something similar and she came down and told me that a plane had crashed on the motorway. At first, because I had no idea which motorway, I was a little annoyed at the football match being interrupted. It was only slightly later when she returned downstairs to say that the plane had come down somewhere on the motorway near Nottingham that I turned the football off and tried to find out what was happening by tuning into the local radio station. Then my deputy phoned up and said 'have you heard the news? A plane has come down at Kegworth'. My initial reaction was, 'bloody hell, City hospital, they're going to have a tough night'. The City Hospital has a major burns unit, and assuming the major injuries would be burns, I immediately thought of them. My second feeling was one of relief, primarily because the Social Services Department had been removed from the Hospital's Major Incident Plan at Queens Medical Centre (QMC), but also because it seemed unlikely that we would be receiving any casualties.

Eventually, it emerged from the news coverage that the plane had not caught fire and that some of the injured were on their way to QMC. I can only describe my feeling then as one of numb panic. Several thoughts rushed through my head – Why had I not challenged the Unit General Manager's decision to take us out of the Hospital's Major Disaster Plan? Should I phone my boss? Should I contact the Emergency Planning Officer? What would our political leaders expect of me and my Department? However, having read previous disaster plans I knew that one of the rules was that I should wait until I was called so I did not set off straight away for the hospital. I had not finished watching the football, so I switched the video back on and at the end of the game listened to the interview with the manager of Sutton United. He was an English teacher, and in his programme notes for the FA Cup tie he'd quoted Kipling's *If*. They read some of it out on television and parts of it haunted me during that first hectic night and have done so ever since in relation to the disaster. I can vividly remember thinking about some of the lines from it later on during the middle of that night when I was in the midst of it all. So I waited for the phone call, but it never came and eventually I thought 'well blow this for a lark I'm actually going to see what the blazes is happening'.

At the hospital

I parked the car and went into the hospital at what must have been eleven o'clock at night. I think by this time most of the casualties had arrived and the medical support services were in full swing. Because I knew the hospital managers, and the location of the Control Room in the disaster plan, I just went straight there and said 'what's going on? Do you need any help? What's the score?' People in the room were already looking stressed, but there was an air of urgent efficiency. It was apparent that the Control Room was becoming flooded by telephone calls from people, particularly from Northern Ireland, inquiring about relatives.

At the time it seemed obvious to me that in order to release key Hospital Managers to continue to organise the treatment of survivors, the handling of these calls needed to be siphoned away from the Control Room. So I said at that point I would be prepared to take over the relatives' enquiries as a Social Services function. I then learnt that there was already a social worker in the hospital who had come to see what she could do, so we went and found her. With the assistance of a man from one of the laboratories, she had already begun to take details of the calls being received from passengers' relatives and friends.

Luckily, we already had some offices very near to where the Control Room was sited, and it seemed obvious that we should set up there. The social worker had begun operating on the back of scraps of paper, anything, just to try and get information written down. At that time, and for some hours to come, as soon as she put down the telephone it would ring again. I was lucky that she had already started a system of taking calls and promising to ring back with information. This was done partly out of necessity, because we did not necessarily have information to hand, and partly because we did not have an administrative system up and running at that stage that could cope with what was happening. Anyway, it was a procedure we stayed with almost for the duration the service was in operation, and I think it helped us keep control.

Setting up a Relatives' Enquiry Service

I made an early mistake here in calling in additional social workers and a senior. In hindsight it would have been better to call in administrative staff first and establish systems. We did this following Hillsborough some months later and it was far more satisfactory. Here we had many administrative problems. We had to hunt for paper and pens. We had no appropriate forms, and even trying to get some photocopying done was a major task. We were based in the duty rooms in the main hospital where even the telephone directories were out of date, and on one occasion I can remember wasting twenty valuable minutes trying to find the STD code to call in a member of staff. We overcame all these difficulties and we obtained the necessary resources, often with remarkable speed, but time can be saved and we learnt an important administrative lesson here.

We had already built up piles of paperwork which was worrying me, for I realised that we needed to establish some sort of administrative order fairly quickly. It was clear that we were now getting several enquiries about the same person and this caused further confusion and anxiety for me as I was worried that if we did not cross reference carefully, there was quite a high chance that we might make a major blunder. It was crucial to our operation that we quickly sort out such problems. That really was the pattern of what was happening. Someone would ring and say 'I am the father of someone who was on that plane, have you got them?' We would say that we hadn't got the information but we would ring them back as soon as we had it, and we would take a number where we could contact them. Then began a process of checking and double-checking our information, logging it and so forth. Of course, we needed to be very clear about the extent of the information we gave out. Although all those who were alive were identified by about five or six o'clock in the morning, we weren't in a position at that stage to release that information. I had taken the decision that when we rang relatives back we would give them a brief factual statement about the condition of the person they were phoning about, and that we would record who had been telephoned, at what time and what had been said. That might be that their relative was 'alive and seriously injured', that they were 'alive and the injuries were superficial', 'that they're currently in theatre', etc. Just something to hang on to. Also, on those occasions when we knew the person being inquired about was dead, because I had been told that responsibility for passing on that information lay with the police, we would have to say that we were sorry but we hadn't yet got information about their relative but could they ring the following number – which would be the police bureau. The decision about passing out information about people who were alive and those who had died was made by the police in consultation with the Hospital Management in the control room.

It must be said though that we were learning these lessons as we went on. We had a brief but intense debate at one point about the issue of confidentiality. This concerned the fact that normally we wouldn't give information out except to next of kin, and here we were not sticking to that. I judged that these events justified a slightly less rigid approach but sensed that not all the Team were happy with that. It felt necessary to be directive about that, so as to take the responsibility away from staff and lay it clearly with me. Slowly we tightened up on that and with the passage of time limited who we would give information to. Around this time one other important aspect of the process of disseminating information occurred to us. We began to ask relatives whether there was somebody within their family who would take responsibility for networking the information when we called back. Thus, one member of the family would spread the information rather than the responsibility for ringing them all back remaining with us. Secondly, there were also a lot of calls from people wanting to offer some information about who was on the plane. I asked the police at some point whether it would be useful for us to collect this information, and was told that collection of information for the purposes of identification was a matter for the police, and was reminded that we had been given numbers to pass on to relatives.

We then started to get some people ringing us back and saying 'we've phoned that number you've given us and it's just the police and we're finding it really difficult to talk to them. They're being very matter-of-fact about it and can we not talk to anyone else'? The social workers were clearly unhappy at what they were hearing and I felt they expected me to do something. Having already been reminded of our role by the police I was loathe to do anything that might result in confrontation. I was also aware of the pressures the police and Control Room staff were under. So, with some courage and a great deal of tact – which is not something I am well known for – I talked once more to the police who were on site and asked whether it would make sense for us to start to collect some of the information needed for identification purposes? The message I got back was, well, yes, if that's happening by all means do, and so we began that process. We weren't really sure of what information to collect. We had no experience of that and we later learnt – within a few hours – that one of the important things to ask about and to record is the jewellery that people wear. It's not necessarily a physical description of the body that was needed – although any significant details may still be important – but the clothes and the jewellery that people wear are crucial in the process of identifying bodies. Not surprisingly, this was very sensitive and very difficult work. We had one member of staff who was involved in ringing back a family several times in order to get a very intimate detailed description of a baby, to help us identify a child that had been admitted to our hospital. We were led to believe that two babies had been on the plane and that only one, the one in our hospital, had been found at that stage. Consequently the relatives this member of staff was telephoning did not know whether their child was alive or dead. That was horrendous – it really was, and that worker spent the best part of an hour sorting it out. She was very traumatised by it, although this did not really manifest itself for some time.

By about nine o'clock in the morning we had managed to return every phone call in the sense that we'd made contact with somebody related to all the patients in Queens Medical Centre. That must have represented well over one hundred telephone calls. Consequently, we were left with offices in a terrible mess – literally papers everywhere. People were tired and it hadn't been easy to get refreshments through the night. Even though we were working in a hospital it had required quite a lot of work just to get some drinks and so on for the staff. In the event I organised it through the Control Room, but it felt like it was a lot of pestering to get it, and I felt quite guilty at having to get the Control Room to organise this when they were under such pressure. Although it can seem a minor point in retrospect, if staff are to be expected to work throughout the night, at the very least they will need a supply of light refreshments.

At this time I was very conscious of what I was doing. There had been enough around in journals about the Zeebrugge Disaster and the like for me to know there was a role for Social Services in the aftermath of disaster. I was also very aware that, on this site, the Social Services Department had no identified function in the Hospital Major Incident Plan, so I'd simply gone in there, spotted a role and grabbed it. That was one factor. I also grabbed it because I thought it was the right role. Looking back, I now know that I was right which, given some of the other mistakes we made, is reassuring. It was a role that wasn't very far removed from the job we do every day of the week and, in my judgement, it was something that needed doing in order to release the Health Service managers to fulfil their role. I also felt, rightly or wrongly, that we would do the telephone work better than other people could do it. The realisation that we would also have another role only came a little later.

The B2 exercise

During that first night I met David Stoter, the Hospital Chaplain, whom I'd worked with previously in a variety of different situations.

He'd been involved in being with the injured when they arrived at the hospital, and by this time we knew this amounted to approximately forty patients. We also knew by then that many of them came from Northern Ireland. Representatives from British Midland - whose plane it was that had crashed – had arrived at some point early in the night, and had let us know that they were intending to lay on a plane to bring relatives over in the morning. I think it was this news which made David Stoter and I realise that the impending arrival of the relatives needed to be planned for and managed. Furthermore, given that the hospital management were rightly concentrating on patient care, we realised that the responsibility for looking after the relatives would probably best be dealt with by a scheme operated jointly by the two of us. We felt there was a danger of the hospital being swamped by relatives, and that we should therefore plan to deal with it.

A brief and somewhat hurried discussion literally took place as we walked along the hospital corridors, and we reached the conclusion that the best approach would be to recommend that the hospital designate a special area to receive the relatives, which could offer some form of protection from press or other interference. I'm not sure whether it was through luck or skill or perhaps both, but we managed very quickly to persuade the Hospital Managers that it was not in their interest to get involved in the planning for the arrival of the relatives. We argued successfully that they should leave that to the social workers and the chaplaincy.

Thus began the growth of what we called the B2 exercise. B2 was the name of a ward in the Hospital which had been emptied and I think cleared for forthcoming re-decoration. We decided to take it and use it as a relatives' area, and began to formulate some plans about how it would operate. We decided that we needed to create a specialist team of workers, and the two main questions facing us were how do we set up such a team and what goes into it? At this point the phones were still going mad and we had to keep them staffed, and that process felt rather separate from actually dealing on a practical basis with the relatives who were going to arrive. I was faced with keeping the two bits of the operation going and wondering how to make sure that the rest of our hospital responsibilities operated in a recognisable form at nine o'clock on Monday morning. Now I took the decision that I would take workers out of each of the available existing teams so as to spread the load. I decided I wanted a team of four social workers and a Senior, who would complement David Stoter's team of chaplains. We also said to the hospital that we needed somebody with nursing experience to help us. The British Midland people arrived and we got them involved as well. They took on the practical tasks of organising accommodation, transport and a whole host of other things, thus freeing our staff for what one might call the 'professional issues'. Having learnt from our overnight experiences we wanted to make sure there were refreshments available for the relatives, so David Stoter and I began to talk to WRVS, because they operated a hospital service anyway in the outpatient area.

The bulk of the relatives had arrived by the early afternoon. We made a decision - not just David Stoter and me, but the team – that we would take a very, very directive stance with the relatives. What happened was that when the coach carrying relatives arrived at Queens, David Stoter went on, I think with some social workers, and basically said 'look, we realise that you're very distressed and that you don't really know what you're going to find. We have an area set up, and a team in place who are going to take you there. We're going to allocate the members of the team to you quite at random (which was the only way we could do it). What's then going to happen is when we find out who you've come to see, we're going to go away and ring the ward up, and find out what's going on. We'll then walk with you to the ward. You'll only be able to stay there a couple of minutes at that point and then you must come back with a social worker or chaplain and then we'll talk about what is going to happen from there'. We hadn't really based this strategy on any model, it was just something that felt right at the time and, with hindsight, we can see that it was really successful.

The message we got back subsequently from relatives that we'd spoken to was that for the first time they actually felt somebody was really in control of the situation. When they went on to the ward there was a mixture of relief, shock, fear, anxiety. A combination of emotions that many couldn't cope with. They didn't feel as though they could break down in front of the person in the bed, because that wasn't the done thing. They actually didn't want to stay on the ward but wouldn't have dared say to their relative that they wanted to leave. With the strategy we adopted they could safely blame us for having to leave. Of course, when they came back on to the B2 Ward there was the most enormous wave of emotion. There were a lot of tears and a lot of distress, and people just pacing up and down. Although this was perhaps one of the lowest moments for the relatives, the Sunday evening one week after the crash was also a very low point. As the time approached that point which would mark the exact time the crash had happened on the previous Sunday, the atmosphere got more and more tense. At this, most likely the worst possible moment, the hospital fire bell was activated, causing very great distress and upset for the survivors, their relatives and the staff Team.

Having dealt with the initial problem of organising the first visits to the survivors from their relatives, having gone through that, the social workers and chaplains were able to ease people back on the wards, at a time and at a pace which they felt was right. That was a matter of their judgement and, by and large, I feel they got it right. People were allowed, given permission to stay for longer periods of time until it didn't become necessary to keep giving permission because they just used common sense. Now what we also said to people was, 'before you go on the ward come and see us and do the same when you come off the ward'. We set that pattern right from the minute they arrived and maintained it. That really worked well and relatives saw B2 as an area they could come to for support, and if they didn't want to be pestered then they wouldn't be, and if they wanted to come and chat then that was fine. We had made provision for refreshments there, so they could just have a drink or a bite to eat. The British Midland people had set up their operation with us and were able to sort out lots of practical arrangements for relatives. The Nursing Sister whom we had requested had also joined the team. She was absolutely crucial to the success of our operation. Although the social workers had all worked in hospitals and had a basic medical knowledge, the fact that most of the survivors needed surgery meant that the relatives had many quite complex questions to ask. The Nursing Sister was able to talk to relatives in a lot of detail about how patients are prepared for theatre, what happens in theatre, what happens after theatre and what to expect in general. She was an absolute brick in that sense. So we had a team that was multi-disciplinary and it was multi-agency. It had the private sector in it, the voluntary sector, social workers, Hospital Chaplains, nurses. We'd never worked together in such close concert before, and we very quickly became a team that gelled.

The next thing to organise were daily debriefings for the staff. I'm not sure we got it absolutely right by any means, but we realised that we needed to exchange information and this became, by default, a form of debriefing for the people who were working on B2. There were two reasons why we needed to do this. First, and particularly in the early stages, there was the simple and practical need to exchange information. Second, because for the first couple of weeks B2 was staffed on a 24-hour basis, there was a need to hand over between shifts. Through both of these practical necessities we established a means of sharing experiences and of providing each other with support.

In terms of the nature of the work that we did with the relatives, I'm not sure if it would be accurate to call it counselling. We certainly did a lot of preparation for counselling. I know that when people went back to their home areas, particularly Northern Ireland, a substantial amount of long-term work was done. I think what we did was to get people in a state to face reality and we gave them information and advice, and permission to do certain things so that they could start to take control of their lives again and plan for the future. At that point onwards they – or at least some of them – will have had counselling.

So here I was with a major crisis, which by that time had got national coverage. I was involved in it and I was determined that as a Department we were going to be involved and that we would do what we had to do well. This seemed to be an opportunity to counteract the view, frequently held by area social workers, that hospital social workers always have the cushy number. I'm quite clear about that in retrospect. I didn't want other people coming in and muscling in on what we were doing, and I kept people at bay. We'd taken a decision about eight or nine o'clock in the morning, David Stoter and I, that we did not want 'outsiders' interfering in what we were doing. I don't think we were precious about it at the time, simply practical. There'd been a lot of vicars, priests, Salvation Army people coming down to the hospital, many of whom were unhelpful. Not all their responses were appropriate, they didn't know the systems, frequently got in the way, and we had to give them the order of the boot in the nicest possible way. We did, however, ask the voluntary sector to undertake some specific tasks for us in the weeks following – but very much under our management.

The relatives' telephone enquiry service we kept going as a separate service for about four days or so. Generally speaking, although we had a shift system, the same workers staffed the phone lines for the duration. On that first Monday the person who'd been handling enquiries about the child was, not surprisingly, in a fair old state. I could see that the other staff were tired and they could probably see I was, but I couldn't recognise it in myself. They were tired and needed a rest, but found it very difficult to let go of it. So, for example, they wanted to meet some of the relatives they had been talking to and consequently there was a sense about of hanging around. There was a real dilemma about actually trying to pace people at that time. Now in hindsight I think we probably should have taken them off that role hours earlier. I now feel that four hours is the most anybody should be expected to cope with that sort of work. Then they need debriefing. I am aware now that we didn't debrief them properly at the end of the session. Some of them went home. Some of them came back to the Department. I mean, looking back it was awful. We organised that poorly. I subsequently picked up all sorts of stories about people going home feeling distressed, not being able to settle, not being able sleep and needing to come back into work, phoning in to ask how things were going.

Similarly, although to a degree I could see what the staff were going through I did not have the ability to see what was happening to me. At the time I felt very strongly that I needed to remain at the Hospital. Partly this was because I felt that I'd made a lot of ground in negotiating many issues with the police and other organisations. I was carrying a terrific amount of information in my head, was aware of the pressure, and so I wasn't entirely confident that if I went, this could be managed by others. In hindsight I'm sure that was wrong. In the end, I felt I should be there. I wanted to be around to help my staff in what had become a difficult situation and, because the adrenalin was flowing, I, like them, couldn't let go.

I think I was as near physically ejected as I could be at about eight or nine o'clock on the Monday night by a couple of colleagues. I went home, and was still very high, both emotionally and aromatically! After a bath and something to eat I sat down in front of the television news. Suddenly there were pictures of the survivors who'd just been names to me during the day. I found that very difficult to cope with. I was really quite cut up about that and found that I couldn't finish watching the news. I just went out into the other room. I think my wife and daughter found that very difficult. They wanted to know what was happening and, because I was a bit numbed by it all, I didn't really want to talk about it. I was actually quite distressed, I was still very high from the emotion of it all and to round it off I was also very, very tired. I know my family found all that very hard. So then it was bed and a very sleepless night. I was back in the Hospital about half past six. Nobody in the Department could believe it. I'm not usually awake until about half past nine and seven coffees.

As staff continued their work on B2 during the first 24 hours of the incident, I was aware of the need to make a decision about whether the small team of social workers should be pulled out of their mainstream duties for the next few days in order to continue their work with the disaster, or whether we should work on the principle that the survivors be filtered on to the main hospital wards and then picked up in the normal way by the social workers covering these units. My preference was to maintain the small team, but I was aware of some rumblings within the Department about this 'elite team', and this made me think very hard about my position. I was encouraged in discussions with other disaster co-ordinators to continue with the idea of a small team, and reassured that they had also experienced similar difficulties when they had been faced with making such decisions. By the Tuesday morning I was firm in my conviction that the specialist team should continue working with these families under the management control of the Senior allocated to the disaster team.

After four or five days non-stop the telephone service was merged with the relatives' area because the number of calls had diminished markedly. The phone lines had continued to be staffed 24 hours a day but towards the latter end of it we'd only one person on it with administrative back-up. The relatives' area ran for the best part of four weeks. Initially it was staffed 24 hours a day, seven days a week, and then latterly seven days a week from approximately 9.00 am until 7.00 or 8.00 pm (to the end of visiting time and half an hour or so beyond). We did close it down before all the patients left. As the number of patients diminished there was less necessity for such a big ward, and by that point anyway the hospital wanted to start getting things back to normal. The relatives used an area by the chapel which was small and discreet, and British Midland carried on their bit nearby.

The needs expressed by relatives were very varied, and one which proved to be important was to visit the crash site. We arranged for two or three groups of relatives to make such a visit. Flowers were arranged for us by British Midland. I can remember watching relatives writing messages, trying to find the right words for notes to loved ones they had not had the chance to say goodbye to. That was a most moving experience. The relatives were escorted by social workers and chaplains and brief services were held there. There were scenes of grief there the like of which few of the Team had ever experienced. Although I was not personally present at each of these visits, there is a poignant story about the last of them. At one point during the brief service the chaplain had to halt proceedings because a British Midland jet was flying low on its final approach to East Midlands airport. After the final prayers were completed there was a strange rumbling noise coming from behind the party of relatives as they stood on the motorway bridge. The group turned round to see the first cars being escorted up the motorway after the M1 had been re-opened. I know this had quite an effect on several people, including members of staff, although for some present this was, I think, a point at which there was clear evidence of things beginning to return to normal.

Checklist

1. Expect that the disaster will be worse than initial reports suggest. People will lose life, friends, property, faith and control. You, your colleagues, friends and relatives are also likely to be affected by your involvement in this work.
2. Having an effective and efficient administrative system is the key to any response you will be organising. Much can be done in advance, and training can be given. It is worth taking time to set this system up before getting the social work task under way. In this context:
   - what have you done to ensure you overcome any problems of communication within the Department and between yourselves and other agencies? There are bound to be difficulties. Can you anticipate some of them?
   - have you got an up-to-date call out list and an agreed system by which staff will be called out? Do you have home telephone numbers and the STD codes? Do staff know where to report, and what is expected of them initially?
   - Do you have a 'disaster box' containing pens, paper, forms, staplers, Blu-Tack®, file covers, up-to-date telephone directories and maps, etc. etc.? Do you know where you will get your photocopying done? (Have you got a call-out number for an engineer if the machine breaks down in the middle of the night?)
   - Have you identified where your service will be sited? Are there enough telephones/points. Are there sufficient electrical sockets for typewriters, computers, a Fax?
   - Will you be using a software package? Are staff familiar with it and with the hardware?
   - Have you supplies of the widely used leaflet 'Coping with a Major Disaster'? If not, can you identify a printer who would print them quickly on, for example, a Sunday night?
   - Do you have adequate identification for staff?
3. Have you considered what space and security you might need for clients and staff?
4. Have you, and the staff, had any disaster-related training? Do staff know the difference between post-trauma reaction and post-traumatic stress disorder, and how to recognise and respond appropriately?
5. Are you clear about the parameters of your Department's role and how it interfaces with others? Are your staff clear?

6. Will you adopt an early strategy of 'taking control'? How will you achieve this?
7. How will staff be selected for the different phases of the work? How will they be briefed and debriefed? Are you clear how longer-term support might be provided for staff in a way that ensures take-up?
8. Have you decided how long staff will work on each shift? Who will ensure that workers go home at the end of their shift, and how will that process be managed? Who will be called upon to work evenings and weekends? How will the issue of 'additional hours payments' be handled?
9. What arrangements have been made to ensure that there will be an adequate supply of refreshments, day and night, for staff?
10. Have you decided about a managerial split between the roles of managing the disaster response, managing staff care, and managing the rest of the Department during the operation?
11. Who will take responsibility for ensuring that communication about the operation is maintained with the staff in your Department who are not directly involved in the disaster work? How will such communication operate?
12. Have you discussed and agreed with the police your respective roles regarding relatives of the deceased, and if/how to collect information to help in identification? Will you adopt an initial policy of 'phone back'? How will you limit initial information? How will you ensure that it is both accurate and kept up-to-date?
13. Where does your managerial support come from? Where will you get personal support from? Are you aware, and are your line-managers aware, that your involvement will continue long after the bulk of the Department's disaster work has ceased?

Note

1. An extended discussion of David Whitham's role in the aftermaths of both the Kegworth/M1 and Hillsborough Disasters can be found in Coping With Tragedy: Managing the Responses to Two Disasters written by him and Tim Newburn. The report is available from Nottinghamshire County Council.

5 Social work in the aftermath of the Zeebrugge Ferry Disaster

Janet Johnston with Liz Beeson

> *Janet Johnston is a senior social worker currently managing the Dover Counselling Centre, a charity which was set up following the closure of the Herald Assistance Unit in June 1988. Prior to the* Herald of Free Enterprise *Disaster, Janet had already established herself in the field of bereavement counselling having been instrumental in setting up the Maidstone branch of CRUSE Bereavement Care.* **Liz Beeson** *is a freelance journalist and a member of the management committee of Maidstone CRUSE. She has been involved on the administrative side of the branch for nine years, during which time she has taken on any role the branch demanded, including that of Treasurer (for eight years), Branch Organiser and, currently, Fundraiser. This chapter examines the social work response to the Zeebrugge disaster, the nature of the 'problems' presented by victims and the difficulties experienced by staff working in the aftermath.*

Introduction

Because of my quite considerable experience of bereavement counselling, when, on Friday the 6th March 1987, the *Herald of Free Enterprise*, whose home port was Dover, capsized just outside the harbour of Zeebrugge, in Belgium, I had a strong feeling that I would be involved. I immediately looked at my caseload of clients, particularly those who were dying, and considered what I would do if I was asked to get involved with the *Herald* disaster aftermath. In fact, it was a full week before I was asked to go, but although I had had time to think about it carefully, in the end I had only twenty-four hours to hand over my caseload.

Following the disaster, the first person who was asked to assess the situation was the Local Manager for Social Services in Dover. Local social workers were at that time staffing a helpline in conjunction with the Police and the Red Cross. After a few days, the Local Manager recruited the Area Training Officer to set up a counselling team, and he contacted me straight away. Both of us had been involved in setting up the Maidstone branch of CRUSE Bereavement Care a few years earlier. The training officer had been given a blunt brief: to get down to the Shipping Office because people 'are shot to pieces'. When we arrived, we found that a welfare team had been mobilised from crew members who had not been on duty on the *Herald* on the fateful night, and they were chasing around trying to give information, good and bad, to the public, the crew and crew relatives. Providing reliable information was far from straightforward, since every ferry had 240 crew working on three different watches, and of the 80 on board that night, 38 had died.

Our first task, therefore, was to support the staff of the shipping company in their work with survivors, the bereaved and their families. The passengers who were on board the *Herald* that night, apart from a handful, were not, of course, locals. Consequently, people from all over the country were converging on Dover looking for relatives and seeking information. Not surprisingly, they went to the shipping company rather than Social Services, because that was where they expected to get information about what had happened.

When we arrived there, we found the staff feeling somewhat helpless; not knowing how to help people if they could not give them information. The very first thing we decided upon was that we needed an oasis in what was, for all practical purposes, organised chaos. We were given rooms in a central, but private, part of the building, and we stuck handwritten notices which said COUNSELLING on our doors. So that was the start of it all. From that point on, at the shipping company offices, and later, in our own building with our own counselling staff – as the Herald Assistance Unit – we made ourselves available at least twelve hours a day, seven days a week for at least the first month; even longer in some cases. And people came. Some came of their own volition and some were 'pushed through the door'. And the real work began.

Recruitment of counselling staff

We now know that of the 542 people on board the *Herald* that night, 193 died, and 349 survived. At the time, we had very inaccurate figures, but we did know that every one of the people on board – passengers and crew – whether they had survived or not, had family or friends who might well turn up in Dover, wanting information and/or needing counselling. Even at a conservative estimate, the potential number of people interested would be in thousands, rather than hundreds. We knew we had a big operation on our hands, but at this early stage we had no real idea of what it would be like. 'Experts' gave us confusing advice about the needs of survivors and we were preoccupied trying to plan the help we wanted to offer both to the bereaved and the survivors.

Recruitment was clearly a priority. For the first weekend there were just the two of us, the Training Officer and myself, but by Monday we had been given a free hand to recruit others. Because both of us had been in the bereavement field for a long time, we had a lot of contacts, so we approached those we thought would fit in with our way of working – become part of the team. We recruited people who shared our understanding of the work we would be doing and would approach it the same way: counsellors from CRUSE and from RELATE; a Psychiatric Nurse and an Art Therapist. We did not have many other social workers. There was a search for suitable people within the Social Services, of course, via the internal circular, but very few applied. I think that was probably because there are not that many social workers who have had counselling experience, or who are comfortable talking about death. We all use interpersonal skills in social work, but that is not the same as being properly aware of the psychological effects of a sudden and horribly violent event such as the *Herald* disaster. I remember our Kent director saying 'Why aren't there more social workers trained in bereavement counselling?'. I had to say to him, 'because the Department doesn't see it as a priority'. We tend not to get involved in hospitals in the Accident and Emergency Departments where sudden death happens every day. Even as a hospital social worker, I found that this was not a priority. My work was more with the elderly, because their beds are always needed, and in that respect I therefore worked more with the ill and dying than with accidental death. But this lack of skill in bereavement counselling is not confined to social workers. I know psychologists who said they would have liked to help but did not have the relevant experience.

Of course, we were looking for very special people. Everyday social skills were in many ways as important as professional skills, but they also had to be people who would be aware of when they were vulnerable to the stress this kind of work inevitably creates, and would learn how to pace themselves. At the same time, they needed to be resilient, too. Besides these basic attributes, they needed to be people who had a knowledge of bereavement; who had been able at some point in their lives to talk through and to listen to incidents of traumatic death. It was, after all, very specialised work. All those involved with the counselling had to support people who, for a period of several weeks after the disaster, had no confirmation of whether or not their relatives had died. It took weeks to lift the wreck and gain access to the bodies, so that even if they felt sure their loved ones were dead, they often did not have access to the body. A lot of them had to make do with seeing a photograph of the body, at most. Helping people to say goodbye to their relatives in these circumstances was a tough task. We had to have counsellors who could cope with seeing the photographs, too, while they were supporting the bereaved as they looked at them. When you think of the disfigurement which violent death causes, you can see some of the difficulties. The three of us had to be sure that we would be able to see this through, as well, because we were responsible for supporting the counsellors. Staff support is vital, as the trauma suffered by those involved in the disaster may be passed on to the counselling staff, who, in their turn, have to cope with it. We therefore found that we needed a very strong support system, and I will be talking more about this and about supervision of staff, at the end of this chapter.

By the Monday after our first weekend of work it had been decided that we would need a building of our own. The Police were giving us information on the people involved in the disaster, and it was clear that we needed to form a unit so that we could liaise with the Police as a team, so that we could become pro-active and develop our own database (which later became 'the Herald System') to enable us to reach out to those people whose names and details were being given to us. Without this base we were out on a limb, and it would have been much more difficult to keep track of the work we were doing, or what we should be doing next. We rented a suitable building near the seafront in Dover from the Harbour Board and this was an absolute godsend, because it provided the right sort of calm atmosphere in which people could come and see us without having to be surrounded by all the business of the shipping company.

By the second week we realised that the task was much bigger than we had initially thought. Because of the country-wide spread of the homes of those affected, we were facing a national task. At this point, the Mental Health Development Officer for Kent and a Psychologist were asked to work for two days a week each at the unit. The Area Manager had by this time relinquished his previous post and taken up the post of General Manager of our Unit, and he realised quite quickly that the training officer and I, with our experience and training in counselling the bereaved, had different philosophies and practices from the two new recruits. We discussed the options open to us and decided that we needed a service that would provide cover for the whole of Great Britain, with a specific, local Dover service to cover south east Kent. The latter was to provide a service for the crew survivors and the bereaved families of crew members who had died, as well as others who worked for the shipping company on the other ferries. It was convenient, therefore, to divide the task between us. Two of us took the south east Kent area, as we had already done preliminary work with a lot of the crew and the rest of the clientele in Dover, and the others took the rest of Great Britain and recruited their own team to do that. We called the two teams the 'Home Team' and the 'Away Team'.

First days and first clients

In the first day or two the Home Team followed the advice given by the Bradford Social Services after the fire at the football stadium there, and had a 'disaster leaflet', *Coping with a Major Personal Crisis*, printed[1]. We ensured that it went out to every crew member who had been on board and every widow or widower or family where somebody had died. A letter was sent with it explaining what we were there for, our aims and our availability. As soon as we had recruited more people we began allocating clients to the available counsellors. We systematically visited each local survivor and the families of every local crew member or passenger who had died. We later extended that, where possible, to those parents of crew members who lived in our area who were not next of kin. Crucially, *we did not wait for people to ask for help from a counsellor*; we wrote telling people that we would call on them on a specific day at a certain time unless they asked us not to do so. The majority accepted a visit, but nevertheless felt that they didn't deserve help – that 'someone else needed help more'. Of the handful who resisted a visit initially, ALL eventually accepted and were extremely grateful that we had given them more than one opportunity to say 'yes' to a visit. I should add that it was very unusual for us to behave pro-actively; without the encouragement of the Bradford team, I'm sure we would not have worked this way.

In the very early days, for the first fortnight, we were working twelve hours a day. Then we acquired a 24-hour phone system from British Telecom, which was really a mixed blessing, as it meant that calls were diverted to our homes – on a rotation system – in the middle of the night. We had this for the rest of the first year.

Whoever was on the night shift had a day shift to cope with as well, as we had no additional staff for refinements. I was living on my own at the time, and declined to take my turn on the night shift because I felt I was too vulnerable to the strength of the emotion that moves people to call in the middle of the night. People who called were always very frightened and vulnerable themselves, and I felt that in my position I had quite enough to cope with, supporting staff and counselling others all day.

The first clients we saw were suffering from a mixture of emotions. For the *Herald* crew in particular, there was a feeling that they were somehow more responsible for passengers' deaths because they were crew. Even though the majority of them could not have done anything to prevent the disaster, they felt they should have known that it could happen. This led to their feeling that they personally had caused the tragedy, that they were *guilty*. This guilt was very raw and all-consuming. Even if they had personally been very brave during the disaster, rescuing people, lifting things they had never known they could lift, climbing things they had never known they could climb, none of them ever thought they had done enough. The crew members seemed to share an unspoken sense of 'group responsibility' and 'humiliation', and it was not only those who had been on board that night who felt this. It seemed to me that potentially every seaman in Dover felt that his life had been spared at the expense of someone else's, particularly those who had swapped shifts for some trivial reason. They did not feel lucky. They felt guilty. Some envied the dead their peace of mind. Most of them could focus on a particular action which they felt they should have done, and for some there were a number of things. They would list things they had realised since the accident, and then somehow turn the blame inwards towards themselves, as if they had not only neglected to prevent it from happening, but had actually caused it, themselves. And it was not only guilt and self-castigation, there was the shame, as well. I had never worked with shame before, and I had to get in touch with my own feelings of shame before I could help anyone else with theirs. Once I had identified and worked through the cause of my own shame (something I thought I had done already – my neglect of my sister who had committed suicide eleven years earlier) then I was in a very good position to help. To add to these emotions and problems which the crew survivors had to overcome, there is also fear – anxiety that the world is a very unsafe place in which to live – and this is also a very difficult thing to adjust to.

All of these emotions were manifest in the first interviews we had with the crew survivors, and at this stage the raw pain was very difficult to deal with, both for the client and for the counsellor. People would howl, sometimes for as long as twenty minutes. It is a particular kind of out-of-control, helpless pain – someone once referred to it as 'wallowing around like a wounded whale', and that, in my experience, is a very good description. Perhaps most people can remember very significant times like that in their own, or their family's lives, but we had it all day, every day. Not surprisingly, it penetrates.

The crew who had died were mostly parents of minors. The impact on their families was therefore devastating. Not only had children lost one parent by death but their remaining parent was suddenly very different. The careful weeks of individual inquests helped many to accept the reality and finality of death. The fact that all the widowed crew knew someone else in the same position meant that mutual support was available from day one, so that although we set up groups for widowed parents and for children these were not used for very long: community and company support for these families was very strong while they went through the painful process of grief. However, not so for the passenger families – those who were not in this close community but were dotted about Great Britain – who felt very isolated and vulnerable. They were recipients of very mixed reactions by their own communities and social services departments – 'What ferry disaster?' one social services duty officer said to one surviving family who sought additional help.

Long-term effects

All the above-mentioned emotions were in play immediately after the disaster, but some of them lasted a very long time. It is my opinion that people are never the same again after an experience like this. Some manage to find a way of life that is OK for them; many lead a very, very different life from the one they had before. It seems that a significant factor in long-term effects of a disaster on people is whether they are able to regain control over their own lives. They tend not to trust people to make decisions for them after something like this. They need to be in control.

They also – still, today, five years later – seem to need to lead more sheltered lives. There are particular things which they find very difficult; going into a crowded pub, for example. Some manage to master it with determination – they go in for a moment, then come out, and then they will try it for two minutes, and come out, and eventually they will manage a whole hour in the pub, but others do not. Reaction to disaster, and the way survivors deal with it, is clearly very varied. Significantly, of the crew members who survived the disaster, only one of them is still at sea.

I mentioned earlier that survivors of disaster have a fear of the world being an unsafe place. One of the phenomena associated with people who are experiencing post-traumatic stress reactions are triggered responses. Thus, things which occurred during the disaster itself, in this case the sound of rushing water, or the high-pitched screaming of frightened people, may trigger the feelings survivors had on board the ship. If someone turned on a tap unexpectedly, for instance, or ran a bath without warning them first, they sometimes found themselves cowering as if they were going through the whole thing again. Hearing children's excited screams while playing in a playground could produce similar reactions. Physical symptoms such as diarrhoea, fainting and tachycardia, are all common during these triggered panic attacks. Some people will have an on-going lifetime affliction, while others will feel they are recovering, and getting along well, until another event happens to remind them, and they find themselves back in the whole thing all over again. We are still – five years on – seeing people at the Dover Counselling Centre who have only just started to have reactions like this. This has implications for Counsellors too, in that some of the team have had reactions.

Effects on staff

Although none of the counsellors, so far as I am aware, has been diagnosed as suffering from post-traumatic stress disorder, all of us were affected, and certainly one or two of the team were exceedingly distressed and continue to this day to be vulnerable to raw pain. This suggests to me that a limit should be placed on the number of clients a counsellor or disaster worker should be exposed to over a given period. For my part, I think that disaster workers should not be involved with more than five families from a disaster, and that they should never (if possible) see more than two or three people in one day. I believe that counsellors' perspective and objectivity is affected by prolonged exposure to pain, anger, guilt, shame, fear and sadness, particularly in such volume. Counsellors' defences begin to weaken if they allow themselves genuinely to share the pain with those they are seeking to help, so they need periods when they can get away from it all and recover before seeing more distressed clients. There were members of the teams, particularly the Away Team, who continued with other jobs whilst doing disaster work. I could not have worked like that. I needed to devote myself to the work completely. Anything else would have been a distraction. On the other hand, I can see that there is perhaps a case to be made for local social workers supporting one family whilst continuing with their ordinary case loads when a disaster strikes. If the family were to need on going support, social workers would, of course, need counselling skills, but for the crisis work, in the first few days after the disaster, this would not be necessary. Their general experience and training would see them through in the short term.

Because of the impact of the work on counsellors, staff support was of prime importance. For myself, I had a personal supervisor outside Dover. This was very significant for me at the time. We were all very caught up in the day-to-day issues, being in Dover, and it was important to have someone from outside the community. The General Manager also gave me some supervision, as did my Team Leader. With my personal supervisor I was allowed the freedom to be angry, sad, confused and proud. From my team supervisor I received care, wisdom and validation for my energetic efforts. From the General Manager's supervision I received trusting permission to talk freely with the world's media, gratitude for taking the responsibility for organising things, and validation within the local community. I had regular access to my home (Maidstone) Area Director who seemed to care about me; the support of a friend who would confront me, another who would just listen and, perhaps most important of all, my three adult children, who carried on as normal, keeping my world in balance. I am not now the same person I was before 6th March 1987. That is a positive statement. The increased sensitivity which this disaster work wrought in me had a part to play in my reconciliation with my husband after four years apart, in 1989.

The Team Leader and I shared the supervision of the Home Team, but in addition, we decided that, as a fundamental backbone to our support mechanism, we needed a group facilitator for the workers who had broad experience of life and death and who could cope with our kind of trauma and emotion. We found a psychotherapist and hired him to run a group for all the Home Team counselling staff at the Herald Assistance Unit. The focus of his group work was to look at what the work was doing to us as individuals. It seemed to work well, for it was the place where I could express my feelings best. Others did not find it as helpful and became angry. We all had differences, of course. The work was very sad and we *needed* to be a bit angry. I think anger is a great motivator: it keeps you going. Oddly, the anger was directed at each other rather than at the disaster, but it did not get out of hand. The management style of our team was a very gentle, caring one, and I think that it did not suit some of our team. When the darker feelings did surface, we weren't quite sure what to do with them.

From our point of view, as counsellors, we were concerned to be in fit mental and emotional condition to be able to support the survivors and their families. Fundamentally, the most important thing is that people really are allowed to talk and that somebody is listening – and I mean *really listening* – and to allow them to validate and *have* validated the overwhelming feelings they have. When people say they feel guilty, most people then try to help by saying 'You don't need to feel guilty', instead of allowing people to own their guilt. Some of the guilt is real, some of it is not, and helping people to sift it is important for their sanity. The same goes for the shame and humiliation – they must discover for themselves what is justifiable and what is not, and resolve it to their own satisfaction. Sometimes this resolution would take the form of doing something practical – for one chap it was designing a safety ladder; for another it was going through the safety manual, and for another it was writing a letter. For the majority, though, it was finding ways of taking control of their own reactions, of getting to understand and integrate these newly acquired emotions into their lives.

In hindsight

It would have been helpful
- to have had some mechanism for prioritising the outreach work.
- to have had more discussions/liaison with other workers in the community.
- to have had structure and training in psychological debriefing for survivors, both in groups and individually.
- to have formally evaluated our work.
- to have had hot meals provided – we neglected our own physical well-being, which did not help our mental well-being.

I would have liked
- to have had access to the 'Experts' panel.
- to have had access to psychiatrists/doctors.
- to have had liaison with local health care workers.
- to have been funded by Social Services to provide a service beyond one year and four months.
- to have been validated by County staff at the time. This did not happen until years later.

In some respects I feel *we did too much*. In other respects I feel *we didn't do enough*.

Maybe we were good enough, but maybe we were not. It is difficult to feel happy with that conclusion.

Notes

1 Reprinted at the end of the book.
2 J. Mitchell (1983) 'When Disaster Strikes – The Critical Incident Stress Debriefing Process'. *Journal of Emergency Medical Services*.
3 Adapted from the work of Atle Dyregrov, Research Centre for Occupational Health & Safety, University of Bergen, Norway.

Figure 5.1 Psychological debriefing (Critical Incident Stress Debriefing)[3]

'Psychological debriefing is a *group* meeting to review the impressions and reactions that survivors, bereaved or helpers experience during or following critical incidents, accidents and disasters. The meeting aims at reducing unnecessary psychological after-effects'.[2] Its aim is to achieve this by:

1. Allowing ventilation of
   a) impressions
   b) reactions
   c) feelings
2. Helping people to make sense of their experiences through
   a) creating better understanding
   b) sharing experiences
3. Utilising group influences to
   a) reduce tension
   b) reduce feelings of being abnormal
   c) share experiences
4. Mobilising resources
   a) Personal resources
   b) Group resources
   c) External resources
5. Preparing for further reactions
   a) a process of normalisation
   b) giving information
   c) an opportunity to arrange further meetings
   d) looking at future needs
      - for the group
      - for the individual
      - for the family

No-one but qualified and experienced counsellors should attempt Psychological Debriefing, as the complexity of reactions is difficult to field even for them. The process should take place within two to three days after the critical incident, and lasts for approximately two hours, repeated after four to five weeks. Ideally, the process should be mandatory for all present at the critical incident.

Figure 5.1 (contd)

Psychological Debriefing Process

1. Introductions
    Introduce self
    Rules

2. Facts
    What happened

3. Thoughts and Impressions
    First thoughts
    Decisions
    Impressions

4. Emotional Reactions
    Questions about thoughts lead to answers about feelings
    What was worst about what happened
    Accepting reactions
    Reactions at the scene and later
    Emotional, physical and cognitive reactions

5. Normalisation
    Commenting on reactions
    Anticipating guidelines
    Advice on helpful coping

6. Future Planning and Coping
    Mobilising support
    Family and children

7. Disengagement
    Summing up
    Follow-up resources

Source: Atle Dyregrov, Research Centre for Occupational Health and Safety University of Bergen.

6 Legal representation after the Kings Cross Fire

Charles Pugh

> **Charles Pugh** is a practising barrister specialising in health and safety and environmental law. His litigation experience includes involvement with the disasters at Kings Cross, Clapham Junction and Manchester airport, and with multi-plaintiff environmental litigation arising out of the Camelford water contamination case. In this chapter he reports some of his experiences of representing families at the official inquiry and the inquests that followed the Kings Cross Fire, and makes recommendations for the conduct of such proceedings in the future.

On 18th November 1987 a disaster occurred at Kings Cross underground station in London. An accumulation of dust and filth on the escalator running track was ignited, most probably by a match which somebody dropped after lighting a cigarette. A small fire began under the escalator and in due time the Fire Brigade was summoned. Within moments of the arrival of the Fire Brigade, and whilst passengers were being shepherded away from the escalator area, a fireball of immense proportions shot up the escalator and burst into the ticket office concourse at its head. Thirty-one people were killed and many more injured, including members of the emergency services who attended. The youngest to die was a child of seven, and the oldest a grandmother in her seventies. Some of the deceased were consumed by the fire but, for the majority, the cause of death was asphyxiation and the bodies themselves were relatively unmarked by the fire.

The Kings Cross disaster had a marked impact on the public. It occurred the year following the tragic overturning of the *Herald of Free Enterprise* in which a similar number of people perished. In the years that followed Kings Cross there was the train disaster at Clapham Junction in which some 35 people died, and then the Hillsborough tragedy. Each of these disasters gave rise to public enquiries and to Coroners' inquests, and each of these disasters received very widespread media attention. The public inquiry into the Kings Cross tragedy was the longest of all the public enquiries, lasting just short of six months, between February and July 1988, and chaired by Desmond Fennell QC. Following the public inquiry the inquests were held and these inquests attracted widespread criticism. A number of members of London Underground staff were called to give evidence as to what happened on the night. This gave the impression to the bereaved relatives present that the Coroner's inquest was a platform for London Underground to seek to exonerate itself from any blame, and this aroused considerable hostility and resentment.

In the report into the Kings Cross tragedy prepared by Desmond Fennell it was noted (p. 161) that to have two separate public inquiries into cases of this sort was neither in the public interest nor in the interest of the bereaved: 'In Scotland the Lord Advocate enjoys the discretionary power to suspend the normal requirement for an inquiry into a sudden death if he is satisfied that the cause of death has been ascertained elsewhere. There is no such discretion in England. Accordingly, I have recommended that the Government should review the requirement in England to hold a separate Coroner's inquest into the cause of death where a public Formal Investigation into the accident has been appointed. In this way unnecessary distress to the relatives and witnesses, with the inevitable additional expense to the public purse, can be avoided' (Cm. 499, 21 October 1988). Unfortunately, this excellent recommendation has not, thus far, borne fruit.

I am a practising barrister and in December 1987 I was instructed by solicitors representing the bereaved and injured to appear at the Coroner's inquest which was opened and then adjourned, and I was then instructed to represent the bereaved and injured at the public inquiry between February and July 1988. Thereafter, I further represented the bereaved and injured at the final hearing of the inquest in July 1988. I had never been involved in a disaster. Fortunately they happen rarely. Unfortunately this means that lawyers do not have a wealth of professional experience to draw upon in order to guide them in the sensitive management and organisation of this kind of work. Thus the Kings Cross 'structure' comprised a team of bereavement counsellors, put in place by Camden to work with those bereaved by the fire and those who survived it; a solicitor representing each bereaved, but with a small steering committee to 'manage' the litigation; two barristers (including myself) who appeared in the public inquiry. This is a complex structure. I personally found the input from the Team Leader of bereavement counsellors to be invaluable, but I was not aware of any formal channel for this input. On the contrary, it arose casually, during informal meetings at the inquiry. Save for this informal contact, and for the support of a solicitor with tremendous experience in industrial/road traffic accident work, I had no resource to draw upon in dealing with the exceptional emotional needs of this group of bereaved persons.

The terms of reference of public enquiries into these disasters are all roughly the same: to find out what happened, why it happened, and to draw from the disaster lessons that can be learned to prevent similar disasters in the future. The remit of the legal team representing the bereaved and injured is to play a full part in this investigation, to deploy the lines of questioning which other public authorities might be nervous of pursuing for fear that the finger of blame might point at themselves, and to ensure that in the morass of technical detail that tends to engulf an inquiry of this kind, the fact that it is about real people whose real lives have been ruined does not get overlooked.

It is difficult to get coherent feedback from a wide and disparate group of bereaved and injured. They come from every type of background, represent every age group, come from all regions of the country, and with many and differing feelings and concerns about what is important, about what should be criticised, about where the blame should lie. Nonetheless one thing I had not anticipated, and my colleagues who have represented similar groups in other disasters have been similarly taken by surprise, was the need of the bereaved relatives to know the most intimate details of what had happened to their loved ones during the final moments of their lives.

It is the natural reaction of someone who has never lost a loved one in a wholly unforeseen disaster of this kind to assume that a relative would not want to know the intimate and horrific details of the last moments of their loved ones' lives. Indeed there is a natural caring instinct to protect these people from those details so as to avoid adding 'unnecessary' distress and suffering to the emotional anguish already in train. In fact my experience after Kings Cross indicates that the vast majority of the bereaved wanted, more than anything else, to know exactly how the deceased spent his or her last moments, where they were travelling to and from and why, who they were with, be it friend or stranger, where they were standing the moment the fireball erupted, who would have been standing next to them, whether death would have come quickly or slowly and in precisely what form.

I spell this out starkly because to a sensitive person lacking all experience of these events there is something seemingly macabre or gory about these details, and something approaching the offensive in someone wanting to satisfy their curiosity concerning them. The pattern was so clear in the group, however, that I was left in no doubt that the filling in of these gaps in knowledge was some crucial part of the emotional healing process, and that any legal proceeding which has as its purpose the concealment or obfuscation of these details perhaps out of some concern for the finer feelings of the bereaved, is utterly ill-conceived. It was apparent to me, incidentally, that there was a temporal development of the feelings of the bereaved in this process. In the first days and weeks people were so shocked that they were numb of all feeling and incapable of articulating a desire for detailed knowledge. It was later in time, generally at least three or four months after the disaster, that relatives started to make the requests for detailed information that I have described above. It is almost a discernible stage in the recovery pattern, and I was left in no doubt that it was a vital one.

The public inquiry was not a good forum in which to conduct this particular exercise, nor, in my judgement, could it be. The public inquiry necessarily is devoted to the 'big' issues arising from examining what happened, why it happened, and the lessons to be learned. In the relentless and necessary pursuit of the answers to these questions it is almost inevitable that the actual deceased become a rather anonymous group and, indeed, the answers to the questions I have set out above in relation to each of the 31 people who died, would not, in point of fact, have helped the Inspector to make any great progress in dealing with his terms of reference, although it would have taken up much time. Furthermore, the hearing of these intimate details in such a public forum with so much media interest and surrounded by a large number of scientific experts, the presence of other interested parties such as the railway unions and the escalator manufacturers and so on, is not an atmosphere in which it feels right to conduct this exercise. We found that we could conduct 'surgeries' with family groups of bereaved in private rooms at the inquiry, in which using the plans and statements of which we necessarily had the fullest understanding, we could take the bereaved through the last moments of their loved ones. This process was followed once again at the subsequent Coroner's inquest, i.e. done privately in a side room, not publicly at the inquest proceedings.

I think that next to the need to know, in the way I have set out above, the next most important feeling of the bereaved person was to see their deceased relative treated as an individual, rather than just as one of the anonymous constituents of the general class of deceased. It is very difficult to tailor legal proceedings to meet this understandable and real feeling. For the reasons I have already given, I do not think that the public inquiry is a proceeding in which this can be done. It can, however, be done at the inquest, by the holding of individual inquests. The inquests after the Kings Cross tragedy were done on a collective basis, that is all 31 inquests were held as one hearing. In other disasters, notably after *Herald* and Clapham though not Hillsborough, the Coroner arranged for 'individual' inquests. Though these may last no more than 40 minutes, those 40 minutes are entirely and exclusively devoted to that particular deceased person, and will bring together all the major details known about the death of that person. Although I am not aware of any study that has been done to follow-up the bereaved in a way which would compare one system with another, I would expect that the 'individual inquests' model is greatly appreciated by the bereaved families.

The feedback from the Kings Cross inquest came in dramatic form. In order to accommodate the large number of bereaved the proceedings were held in the Shaw Theatre on the Euston Road, the proceedings taking place on the stage, whilst the bereaved and public sat where the audience would normally sit. Initially the bereaved were unrepresented. Legal Aid is not available for legal representation at Coroners' inquests. When witnesses were called who gave accounts of the events of the night in question without any cross examination, the bereaved were extremely upset and sought, and obtained, emergency representation (for which London Underground Limited ultimately paid). In the event some of the ground which had already been covered exhaustively by the public inquiry was gone over again to the satisfaction of nobody.

Once again, in the context of a disaster of this kind, the lawyers are generally unprepared and inexperienced in dealing with media interest which centres upon the bereaved families that they represent. When, for example, lawyers represent a widow in an industrial accident case, it is a question of sitting down and explaining the legal principles concerning the availability of compensation, the need to prove fault on the part of the employer and so on. This is not undertaken in the context of media interest. Following a disaster, by contrast, the media are interested in what the bereaved relatives have to say, and the relatives need assistance in handling this new situation, but it is not assistance which lawyers are trained to give. Following the Kings Cross disaster, bereavement counsellors were appointed to assist the families and no doubt did a great deal of valuable work. Here again, lawyers have not been used to working with bereavement counsellors and were lacking in experience as to how they should collaborate with them, there being no guidance to refer to on this subject. Although there was co-operation between lawyers and bereavement counsellors, I felt that we did not know enough about the sort of work the other was trying to do to get the best out of the relationship from the point of view of helping the people concerned. I think that a co-ordinated approach with regular meetings between the counsellors and lawyers would be a considerable improvement.

Looking to the future, I consider that the public inquiry and the Coroner's inquest do have separate tasks, but that there should be individual inquests focusing on and tailored to the assembly of and exposition of all known information on the particular deceased. I think that the 'need to know' should be recognised and that the legal team should work closely with the bereavement counselling team in ensuring that the need to know is satisfied in the most appropriate way. Concealment, obfuscation, and downright lying to protect a person's feelings, however well intentioned, simply will not do. That said, in order to pass on to bereaved people details which are graphic and deeply shocking is something which the lawyer can best do with the help of really experienced bereavement counsellors, and there should be much closer co-operation between these professions.

7 The role of the general practitioner in the aftermath of the Lockerbie Disaster[1]

Margaret Mitchell

> Dr Margaret Mitchell is a Reader in Psychology at Glasgow Caledonian University. She is a chartered psychologist and a member of the International Society for Traumatic Stress Studies, and has studied the effects of disaster on several groups including the police and other emergency personnel. General practitioners potentially have a vital role to play in the weeks and months after disasters, particularly as an interface between those affected by disaster and other services. In this chapter, Margaret Mitchell looks at the example of Lockerbie and examines the role of local doctors.

This chapter is based on interviews with the seven general practitioners in the Lockerbie and Lochmaben practices. The interviews were conducted in September 1991, 33 months after the disaster, and reflect not only the immediate experiences of the local medical practitioners but also developments and recovery in the community since that time.

Background

In and around Lockerbie there are two main medical practices, one in Lockerbie itself and the other about three miles distant at Lochmaben. In total there are seven general practitioners. In smaller rural communities the general practice acts as a hub for the whole area, and what was interesting about speaking to the local doctors was their role as 'participant observers'. These individuals not only had their own experiences of the disaster to deal with, but also the task of treating and continuing to provide support for people, most of whom they had known prior to the event. They were, therefore, in an unusually informed position to judge how particular individuals had been affected, and this provided valuable insights into people's vulnerability to emotional disorder. All of the doctors were there on the night and remarked how helpful it was in their later treatment of the patients to have experienced the disaster first hand. Some searched for possible survivors, although it was quickly realised that the search was fruitless and others treated the very few residents of the town who were injured.

On the night of the disaster in anticipation of hundreds of serious injuries the area was placed under full medical alert. This, of course, turned out to be not necessary. It is a phenomenon of Lockerbie that, despite the nature of the disaster in which all the occupants of the Pan-Am jet lost their lives, mercifully few on the ground were killed or injured. The community was considered fortunate (in a relative sense) that large areas of the town had not been consumed by fire, and that the damage had been contained. Although not evident in the immediate aftermath, it became obvious that the damage to the community was psychological rather than physical, and was profound, far-reaching and insidious. The terrorist attack on the plane which left hundreds of families all over the world tragically bereaved, also emotionally damaged a whole community. Doctors are expected to be knowledgeable in not only a wide range of physical and psychological disorders: this expectation is, if anything, increased in traditional practices such as are found in rural communities like Lockerbie. Although this is now changing to a degree, a general practitioner's knowledge of emotional disorders is obtained on the job and very little formal training is provided to help them recognise and appropriately treat such disorders.

In the days, weeks and even months after the disaster, patients were coming in to the surgery reporting the same sorts of symptoms: sleeplessness, losing interest in work or previously enjoyed activities, being disinterested in the family, being easily startled, being scared of the dark or scared of loud noises. Children's behaviour regressed: they didn't want to sleep alone, or they wet the bed or cried and misbehaved. While all of these were seen as quite normal reactions to an abnormal event, it was when they persisted or were extreme or intense, that both the doctor and the patient agreed that 'something was wrong'.

Of the greatest significance in the doctor's understanding of the longer-term effects of disaster and trauma was their provision of medical reports for those claiming compensation. In the natural history of the aftermath of the disaster this had two major effects. First, to assist the doctors conduct systematic interviews the solicitors, on advice from psychologists and psychiatrists, furnished them with literature on post-traumatic stress disorder (PTSD), a condition about which few of them had heard, and of which none had working knowledge. All the doctors say that this material allowed them to make sense of their patients' symptoms, and to recognise and understand them as classic signs of PTSD.

Second, claiming compensation required a visit to the doctor by people who, the doctors doubt, would have come if left to their own devices. Lockerbie has often been described as a stoical community, and there was a strong sense amongst the residents in the community that they should be able to deal with their experiences on their own and not make a fuss. There is little doubt in the doctors' minds that they would have suffered in silence for as long as possible, as indeed many tried to do.

While some of the doctors acknowledged having had a degree of scepticism regarding the legitimacy of psychological illness compared with identifiable physical illnesses, they are all now 'in no doubt' that PTSD exists as a valid disease entity.

The night of the disaster

On the night of the 21st December the doctors in Lockerbie were, like everyone else, just going about their business: finishing a surgery, having an evening meal or enjoying some social activity. One who was at home at Lochmaben said: 'We heard a tremendous noise like a lorry crashing outside the house, but it went on and on and I realised it was going on far too long to be thunder. We looked out and saw a huge mushroom of flame. At first I thought there had been an explosion at one of the nearby factories. Later, when I realised how far away it was – about four miles – I was staggered to think how big it must have been. We drove straight to Lockerbie and I heard (another doctor) on the ambulance radio, so by the time we got there I knew what was going on'.

One of the doctors from the Lockerbie practice had assumed the role of Principal Medical Officer and co-ordinated the local medical response until about six the following morning. He and another doctor were on the scene to ascertain what emergency medical care was required, and were joined by others from both practices. 'There were a lot of people wandering about. Some of them had cuts and bruises and many looked quite dazed. It was decided that we should set up a casualty clearing station at the surgery for the 'walking wounded'. With the help of the nurses who had arrived, I prepared each of the rooms for emergencies'.

'We were expecting lots of people, but only about two dozen

came in who actually needed medical attention for wounds. Others gravitated towards the surgery and we spent a lot of time just reassuring people, which was what they needed. The surgery became a place where townspeople came to ask if we had seen someone. One or two of the people who had relatives killed were going from place to place looking for them. We worked like this until about one or two in the morning.'

'Anything we wanted was instantly there. For example, we had no water and no light and we wondered how to treat people with burns. We contacted the local chemist and he brought us bottles of distilled water and cream for burns. So many people offered help. I must have had a couple of dozen general practitioners and the same again of nurses who wanted to help. A lot of the time we spent explaining there was nothing to do, but thanks anyway. I think that caused quite a lot of frustration in people who wanted to help and that was very difficult'. A number of doctors also helped with the searches: 'One or two of us were allocated to each search team and we were sent to different areas. We went with some local people who knew the lie of the land and a couple of police officers in radio contact with the police office. We searched from midnight until three the following morning, tagging any bodies we found. In retrospect, this could easily have been done the next day. It was dark and it was wet and the bodies could not be moved anyway until the forensic scientists had a chance to take an overall view. I suppose there was always the hope we would come across someone alive, but when I look back now I think all this activity was pointless. We all felt we should be doing something there and then, and perhaps it was more for our benefit that we did this; to make us feel that we were doing something useful. We wanted to save lives and at that time we all felt very inadequate.'

'We thought it important that the local doctors should do this work because the disaster was within our own community. We didn't want the local people to think we couldn't cope. Perhaps it is irrational, but we didn't want the police surgeons from Glasgow coming. It is true that some friction developed and we felt a bit annoyed at some of the outside medical officers. Emotions were high and I personally felt very angry at what I saw as disorganisation on the night and for a few days afterwards. The Town Hall and the centre of Lockerbie were absolutely jam-packed with emergency personnel, many of whom were standing around. I was angry at the influx of emergency personnel from across the country, and largely they were just getting in everyone else's way. They probably weren't, but that's what it felt like. I suppose we felt that they were intruding on our patch. Yet I quite accept that if it had all happened in a different way, there would have been huge numbers of casualties.'

'The next day we were back to normal at the surgery, although it was very quiet. The following day, on Christmas Eve, I was asked to take part in the recovery operations in Rosebank Crescent, where the army were digging out bodies from the rubble. We were asked to certify the death medically – although it was obvious – and also to identify any body parts as required. The police were taking photographs and films of each body and each part, and everything was tagged after identification by a doctor. I cannot say it was a pleasant experience to spend five hours at the crash site.'

The next few weeks and months

'In the first few weeks afterwards the surgeries just went to pot. People seemed to have an overwhelming need to express what they had experienced, about their involvement in the disaster and how they were feeling. People talked about little else either out in the street or in the surgery, and this went on for many, many weeks'. The doctors recall that people, thinking that the surgeries would be inundated with patients, did not come for the first few weeks afterwards. 'Then patients came forward with physical complaints rather than any psychological disturbance. But it was quite clear in my mind as well as in the patient's mind what had happened – for example, an asthmatic who suddenly got worse on the night of the disaster and had not improved since. One or two angina cases who got worse. These sort of things that were clearly anxiety related.'

The practitioners at Lochmaben tried to contact as many people as possible within their practice who had been near the devastated areas. 'Two of the families in the practice escaped through flames to get out of their houses that were badly damaged by fire. They were being seen anyway but we arranged to see others too. To be honest, I didn't know what I was looking for at the time. I thought I would be able to identify problems just by talking to them. I think that you could say I went in blind, asking "How are you? or Have you any problems?". You know general questions. And they would say things like, "I am fine, maybe wee Jimmy isn't sleeping, but we are managing fine. Considering what happened I'm fine."

'We both, that is the patient and I, thought that almost any reaction was normal to this devastating event. Many, at least superficially, were their normal selves and were coping, saying "thanks I don't need any help". And I was duped by that. I know that I missed some of the more severe cases at the beginning. A depressed patient says "I can't cope, I am crying a lot, I feel really depressed". They furnish you with symptoms and I thought that these, as I now know it, traumatised patients might supply me with the symptoms which I could then interpret in terms of anxiety or other emotional disturbance. But the patients were not forthcoming, which is unusual.'

'Listening seemed to be essential and also what the patients wanted. A lot of time was spent listening, and doing nothing else. Often I found myself listening to the same person giving their same story again and again. Possibly they may have needed or received a more structured interview from other people. But for my part just listening seemed to benefit them. People suffer catastrophes every week: sudden death, terrible illness and so on. On a one-to-one basis the skills which were being asked for were really no different than those required in these other situations. Possibly the only difference was that the distressing event was so unexpected and so major, and affected so many people that it was different in its intensity.'

As it became obvious that some patients were not improving, or when they showed signs of a more depressive illness, some doctors began to think of drug therapy. 'I wasn't rushing to prescribe before about six months after the disaster. In fact ... even six months ... I've just plucked that figure out of the air ... it may have been more like a year. Overall I don't think I prescribed a lot of drugs. Mostly it depended on my gut feeling whether I was going to be able to help this patient by any other means, or if my listening was not a productive use of either my or the patient's time.' It is evident that the amount of time that general practitioners were able to spend with each patient was limited. The difficulties in managing this in the few months of surgeries after the event is clear from the doctor's expressed feelings of being under time pressure, caught between wanting to listen to the patient, aware that the patient may not have told their story to anyone else, as well as being conscious of the next patient waiting. All the doctors alluded to the possibility of a referral to the community psychiatric nurse, although it not clear how many referrals were actually made. In the Lockerbie practice in particular, back-up assistance was available from a team of psychiatric nurses and a psychiatric consultant. There seems to have been very limited contact with the social work department, either in the form of referrals to or from them.

In fact the timing of the decision to take other action, make a referral, or to prescribe, really seemed to depend on the doctor's perception of the individual patient. They were more alert to the possibility that patients who had a history of psychological disturbance might need help earlier. Indeed all the doctors had expected that patients with 'previous neurotic type illnesses' would be rushing to see them. 'But in practice, it wasn't like that at all. We were surprised that some people who previously had coped with whatever life threw in their direction, were sometimes the most distressed and disturbed. And others whom you would have thought would have been very shaken by their experiences did not appear to be all that upset.'

In Spring 1989 the doctors in both practices were asked for medical reports for compensation claims. Being provided with information on PTSD 'gave us a checklist of things to look for. We were given articles about what to look for in patients with PTSD – and suddenly everything became clear – I had a clinical condition with symptoms I could look for. It formalised the way we were looking at patients and their reactions, and brought some organisation into the medical and clinical assessments. So, instead of saying to a patient "how are you today, are you alright?", I was able to say "are you sleeping?", or specify some other typical symptom. We did hundreds of these legal reports and so we got more perceptive at picking up the symptoms of trauma.' It was then realised that people, who were now being encouraged to discuss their symptoms, had been under-reporting their symptoms and, on the whole, were being very stoical. Amazement was expressed 'what people were putting up with and what they themselves accepted as an understandable reaction, like very extreme nightmares, or having the symptoms for six months. Their story would come out bit by bit. Often in describing it they would be quite upset and were in tears. Some of them I wanted to see again or even put on anti-depressant treatment, and it worries me because I doubt if they would have come to see me off their own bat.'

'I think they felt guilty and didn't want to be bothering the doctors with something that they thought they should be coping with themselves. The people were quite reticent about their symptoms because they felt that everyone must feel the same way and they had no right or need greater than the next person. They only came because they were told to by the lawyer. Some came and were embarrassed because they thought they should be getting "over it by now". They had compared themselves with their friends and neighbours and thought "I know everyone felt like this to begin with, but my neighbours are better now and I am not". There were a lot like that. It wasn't a question of whether there was something there. Here, quite clearly, was a very distressed person, and whatever label you gave it there was a clear cause. Some people couldn't go out of the house, or go shopping, or go to the end of the street, or they couldn't get on a bus, or they couldn't go out in the dark, or they couldn't drive in the dark, they were so afraid of loud noises that if a car passed them they just leapt to the side – they couldn't hear the noise of trains going by without getting frightened. And other people, often in the same family – which made me wonder how on earth they managed to cope – would have the total opposite reaction and they couldn't stay in the house. They walked round and round outside. They really had quite florid descriptions and you wondered how these people could live together. It was very sad and a lot of marriages went through a lot of stress.' There was often 'a feeling of chronic low drive and of not getting on with things. Others came along with nihilistic feelings and general emotional blunting. Several were sent by partners as being impossible to live with.'

'The older people were those who went on the longest without getting help. The older men had particular difficulty coming to terms with what they regarded as their own weakness. They thought they should be able to cope, and when eventually they found they couldn't, they came for help in quite a bad state. Of course, the older people had other problems to deal with as well. Some remembered experiences during the war which kept recurring in dreams. Others were also deeply upset at the other disasters which happened and were on the news. Each one that happened brought in a spate of people for the following few days.' The fact that the doctors were from the same locality was seen as a distinct advantage. 'Being from the area helped vastly. A lot of people felt the need to check if you had been there before they started to tell you. They almost wanted to establish a sort of fellow feeling before they would really let all their feelings out to you. And quite apart from that it would have been very hard to really understand what happened and how people might feel if I hadn't seen what they were talking about. It is obviously a very difficult thing to describe and if you have a very vivid and graphic picture of what it looked like yourself, and know that everyone else in Lockerbie has that too, then you don't have to try and describe it. You can then get on with talking about how they felt.'

In retrospect

It is an unusual situation for general practitioners to be faced with several instances of the same disorder, and it is even more unusual for them to be faced with widespread psychological distress in the community they serve. One of the most prevalent preconceptions in understanding psychological disorder is that only certain groups or types of people are prone to emotional distress. When general practitioners see only one or two cases of psychological illness it is easy to look for causes in the people's characters or in their environment. In this case the effects of the disaster were so far-reaching that such preconceptions were challenged. The doctors were struck by the degree to which people from all walks of life were emotionally affected. It was a surprise, for instance, that working males – who as a group are considered more resilient and good at coping – were also traumatised. Eventually, as the general practitioners dealt with more and more of the residents of Lockerbie, the striking feature was that it was not necessarily the weak or those who had had previous psychological disturbance who were affected, but that everyone was to some degree.

'From my experience, I can't identify any themes or any typical vulnerable people, and I don't know how this can be explained. I have often wondered . . . I've looked at families and we have perhaps had one member – perhaps an adult – badly affected, a child who has been partly affected, perhaps another child who has not been affected, and another adult who has not been affected either. So within one household, where everyone has had the same experience, you have all these different reactions. I can't explain why that would be.'

It was clear at the outset that both patient and doctor thought the reactions were normal responses to an abnormal event, and that they would get over it in time. 'Most patients have, at least superficially, without a great deal of expert help, come to terms with what happened and are functioning normally. Thinking back now, quite a lot did not have what you might call a straight progression to feeling better. People seemed to get over things and were fine initially. With the darker nights and the anniversary they noticed their distress returning. At the time of the anniversary new patients – people we had not seen before – were presenting. The second anniversary seemed a bit better, but certainly at the first anniversary we saw new patients as well as regression in the well-being of some who had been a bit wobbly earlier but who had been coping. I daresay that if you started questioning and probing there may be something of an iceberg. There may be some symptoms lurking away there, but most people are coping now, and there are no new presentations.'

'Its the old story of letting sleeping dogs lie. People are getting on with their lives and they have sort of blanked out their experiences. That is until suddenly there is a big reminder like anniversaries. And it also happened with the news coverage of the Fatal Accident Inquiry. A lot of patients who were in to see me for other complaints would spontaneously say they wished they didn't have to hear all that again. It brought back all sorts of unpleasant feelings that they were quite happy to forget about. They had been doing quite a good job of putting it all behind them.' Now over four years after the event, doctors in both practices agree that the community is generally well on its way to a good recovery: 'The first major watershed was probably the first anniversary. And I think generally the community and the people were pleased to see that behind them. They needed the time, I think, for life to get back to normal, and also the withdrawal out of the limelight and away from the intense media involvement there was to start with, and a return to normality. Just like it can sometimes take a person months or years to get over the loss of a close friend or relative, it will take some individuals a long time to get over their experiences in the disaster.'

'Today a lot of people in the community accept minor symptoms as a part of life. It is accepted as normal that children are afraid of the dark, or don't want to sleep on their own. And people just accept this as being the result of the disaster. Many don't like the sound of helicopters and planes passing over, or they might watch the flight of an aircraft to see if it comes down.' In a general sense the doctor's experiences of the disaster have led them to consider psychological factors more in consultations. This may also be helped by the patients being more accepting of the effects of psychological factors, and their influence on physical illness. One of the doctors feels that the disaster has given people 'permission' to speak about psychological issues which they would never have done previously. Their experiences also have implications and lessons for the aftermath of disaster in other communities.

Whatever interactions one would expect between the different support and service agencies in the community, one might expect this to be enhanced in a post-disaster phase. It is striking from the reports of the doctors' experiences, however, that they appear to have had very little contact with the social workers or with the social work department. Possibly this is even more unexpected when one considers that their work was in some respects – particularly the emphasis on listening over and over again to people's accounts – similar to that usually undertaken by social workers in the aftermath of disaster. Overall there appears to have been very little contact with other professionals in the community, other than the regular link which was established to a local psychiatrist and to the psychiatric nurse attached to the Lockerbie surgery.

The interesting exception to this was the contact with the lawyers. This was initiated by the lawyers themselves in their requiring the patients to attend the doctor for a medical report. As is evident, this appeared to be vital in bringing forward many people who otherwise might not have received help because of their reticence about complaining.

It is evident that community general practitioners have an important role to play in the aftermath of major tragedies, primarily because of the position in the community they are perceived to hold by members of that community. Doctors also have the facility to educate the community in terms of anticipated reactions to disaster and 'normal' symptoms. In addition, they can refer patients on to other psychiatric and psychological supports when this is required. Clearly there is much that the general practitioner can offer to and contribute to other professionals involved in disaster response work, and in turn there is a great deal that doctors can learn from them. There would be considerable benefits for everyone concerned if the lessons learned at Lockerbie could lead to the integration of the general practitioner into a far wider helping response in such unusual circumstances in which the responsibility for 'informed listening' and support could be shared.

Notes

1 The research reported in this chapter was supported by the Lockerbie Air Disaster Trust. The author sincerely thanks Drs Adam, Frost, Hill, Longmore, McQueen, Sloan and Taylor for their co-operation.

8 Mental health and social services: working together after Clapham

Carolyn Selley

> *Dr Carolyn Selley* trained as a psychiatrist in Birmingham and later as a psychotherapist in the Southampton Department of Psychotherapy where she now works as a consultant. She headed the Clapham Support service after the Clapham rail accident and has conducted a follow-up study of survivors of this disaster. In this chapter she examines the role and work of the psychotherapist and considers the difficulties of working alongside other agencies in the aftermath of disaster.

Although a week may be a long time in politics, four years can be a short time in disaster work. As I write, the fourth anniversary of the Clapham rail accident is a month ahead. There are still survivors living with their injuries, compensation claims unsettled and court hearings planned into next year. Leaving aside these practicalities, the news this morning tells me that there are train delays at Clapham Junction because of a signal fault (the cause of the accident those many months ago). There are differences between now and then. My husband is on the commuter train this morning but he wasn't four years ago. I am a survivor of disaster work but I was not on duty on the 12th December 1988 when it happened.

Two commuter trains travelling through Dorset, Hampshire and Surrey were involved, carrying over 700 passengers. Thirty-five people died and around 120 were treated in hospital, a third of these requiring inpatient treatment, and including very serious injuries leading to long-term disability. When the tragedy occurs, however, such information is not available and only later does it gradually seep through. Frequently, the media is the only source of information at first, although most of us would gladly 'shoot' this sometimes thoughtless messenger. It can't be right that a child first learns of his father's injury during children's TV when the screen shows him carried from the train. I was a senior registrar in a psychotherapy department working with much individual trauma, but work with victims of mass disaster was new to me. There were few immediate colleagues with such experience. The meeting we held the day after the accident, convened within the Department of Psychiatry in the hospital, was attended by a local social services manager, in addition to staff from many disciplines within the mental health services. The details of the event which had occurred some 60 miles away were still unclear, and it appeared very remote from the heavy demands of our everyday work. It was decided at the meeting that I should be responsible for co-ordinating the response of Southampton's Department of Psychiatry and that funds would be sought from Wessex Regional Health Authority for five consultant sessions a week for me plus secretarial support.

The major areas of work

'Clapham Support', a service for all Clapham victims in the region, was set up immediately and National Health Service funding released for just two consultant sessions a week, two months later. The Clapham Accident Trust Fund and British Rail were approached but, for different reasons, declined not to contribute. However, two years later, British Rail, in response to a survivor's request, provided a substantial grant.

Disaster strikes individuals at random. If you turned to the right rather than the left to look for a seat on entering the first carriages of the Bournemouth train that day, you were more likely to survive. In the same way it is often fortuitous who takes on disaster work. Who can offer the extra time? It happened to be me. I had worked for many years alongside therapists from other disciplines such as social work, psychology and nursing and, consequently, felt open-minded about working with new colleagues in order to help Clapham survivors. However, I was not prepared for the problems that lay ahead. In the first few weeks there was great emphasis from the media on the early disaster work. Ironically, this diverted attention from the importance of planning long-term services. By the third day a GP, familiar with the Department of Psychotherapy, asked for advice and enquired about referral to us of two people who had been on the train. In response to this and other requests, a helpline was set up by the Clapham

Support service during the first week to provide information about the help that was available.

In the absence of any other group in the area, we started our weekly Clapham Group on Day 10, with eight people coming on the first occasion. This group ran for two and a half years with some people leaving and others joining as time went on. In all, 28 people became members. It fulfilled different roles at different times, initially providing a meeting place for support and venting feelings, and later running as a dynamic psychotherapy group, but always focusing on the accident experience. There were always two facilitators in the group, and supervision and support for them took place after each session. Facilitators' awareness of group processes and experience of group work was vital to its success. All survivors who were experiencing severe problems six months after the accident were given the opportunity for an individual meeting with me for an often lengthy overall review. I suggest that this should be offered earlier, after three months, if time allows. This gives an opportunity for a full appraisal of the problem, taking into account individual factors operating, and allows a plan for future help to be discussed.

GPs, some social workers and staff at St George's hospital where the injured were taken, referred to me any survivors who developed post-traumatic stress symptoms. There were also some self-referrals. Most wished to join the weekly Clapham Group, but others requested or were deemed to be more appropriate for individual work. Those seen individually were sometimes suitable for brief focal psychodynamic therapy. Again, support and supervision followed each individual session of therapy. Some of those who were bereaved, and whose grief was not progressing, were referred to me by local GPs. Some were taken on for individual grief therapy. Others attended a Clapham support group specifically for the bereaved. Both I and social workers from another area faced a common problem. We were uncertain about the merits of separating the bereaved and passengers into two groups. There were pros and cons. Thus, for example, the strength of some of the bereaved in facing their pain was a great help to some passengers. Also survivors were the only ones who could tell the bereaved of their loved one's last journey[1]. For the passengers who had seen the uncollected cars in the station car park when they returned home after the accident, it was valuable and eased their guilt to realise they could help. The personal strengths of the two groups appeared additive at times. However, there were also differences in the work that was required for them and I separated them after the first few group sessions. Approval for this was expressed by some but was not universal. In an ideal world a regular common informal forum for all, separate from the special groups, would have been helpful.

We were involved in a wide variety of other activities. As time went on the people attending the weekly Clapham Group wanted to see the train wreckage. This had to be negotiated with British Rail and the British Transport Police. I arranged for a hospital social worker to accompany them. The memorial ceremony information was disseminated through the Clapham Support service. One volunteer produced a newsletter. I started an outreach programme nine months after the accident. This involved contacting all the bereaved living locally and informing them of available services. In addition, all injured or travelling in the carriages most affected, not already in touch with the service, were written to. At times I felt as if I had two heads: therapist and social worker. Others acted as if I had! However, those experienced social workers, psychologists, doctors and nurses I knew from before the accident worked with me in the Clapham Group, with the bereaved and on helplines, in a most skilled and selfless way.

I arranged disaster worker meetings for the geographically scattered group of Clapham workers. Practicalities such as travelling time prevented frequent meetings, but we kept in touch by phone, and each area made local arrangements for more regular support. Every six months I arranged a review day for all workers, which gave a further opportunity for support and discussion of problems. Alongside the service for Clapham victims I started a research project to assess the long-term effects of the accident on survivors. It included those injured and those who escaped physically unharmed. This work allowed further outreach. Those showing high symptom scores could be identified and recontacted. By the second year there were many who had coped with the event but others were still in difficulties. The resilience of individuals in the face of this experience increased my respect for those who have to struggle with emotional trauma of this kind.

Working with other agencies

In the first few days I publicised the service to relevant hospitals and GPs throughout Wessex and I also used the newspapers, television and radio. There was intense interest in survivors and we required good hospital security in the first few weeks. A request for televising of a group session was turned down as were several enquiries about the possibility of conducting individual interviews with those involved. Our service had to be seen to be completely confidential, and I had to be seen not to be associated with the worst excesses of the media. At times I felt that undue pressure was put on me by reporters to give names of survivors, occasionally accompanied by the threat that the alternative was for the journalist to turn up on someone's doorstep indiscriminately. However, there was a plus side to the media involvement. We received the sympathetic co-operation of some local and national press and we were also able to use newspaper cuttings in some of the group sessions to help with the understanding of what had happened, and to encourage the expression of feelings in a constructive way.

There was much liaison with other agencies. In the first few days I found that at least four social services departments were involved in providing help for those affected, and mine was one of three hospitals taking a prominent role. One of these, a privately run hospital, was contracted to supply services to employees of a large financial concern who had been on the train. However, the view of the passengers as affluent and successful was not always accurate, particularly when the symptoms of post-traumatic stress led to employment insecurity and redundancy.

Much of the action, early after the accident took place around St George's Hospital in Tooting where most of the injured were taken. The liaison psychiatrist and psychotherapy department there immediately made efforts to establish contact with all casualties and arrange debriefing sessions for staff. The 'Clapham Support' service would have required a much larger number of staff if these tasks had been necessary locally. We were able to liaise with St George's and there was no problem in my obtaining names from medical colleagues for outreach. As a result of telephone conversations with medical staff there, I became aware of two planned follow-up group sessions in London for their patients. This liaison continued for six months, after which a social worker colleague and I met in Southampton with St George's staff to discuss the transfer to us of care for some survivors and to mark the stand-down of the St George's operation. We all learnt new information that day about the experience of the accident and gained from further piecing the jigsaw together both for ourselves and those involved. I was alerted to people who were not yet asking for help but who were thought to be in excessive distress. As workers, we also gained from exchanging views and talking of the feelings invoked in us.

I found problems arose for me in communicating with other services and in the varying approaches taken by the different social services departments. Some problems were ones common to us all irrespective of our background. About a month after the accident, and after much fruitful phone communication, I attended a meeting of social services representatives, and was able to express my wish to work closely with social workers. However, this was the last such meeting organised by social services management for all the areas involved. After this, and for the following two and a half years, the different social services departments managed their responses separately and in different ways.

The 'specialness' of survivors was beginning to emerge. Could they be stigmatised by contact with social services or psychiatry? Were they patients, clients or what? Should we keep our usual notes? These were questions I and others were asking. One social services department issued a statement to its staff that individuals who were seen as a result of the accident would have their records destroyed if there had been no further contact after six months. Other departments decided, like me, to keep their records. One social services department that took responsibility for the bereaved, visited within 24 hours but, if not asked to call again, did not follow-up. However, this meant that nine months later the names of the bereaved were not easily available when a group was set up by two social workers working with me. Eventually British Transport Police provided me with the names, and 12 took up the offer of a group meeting. In other departments, where workers kept in touch with the bereaved well into the second year, they tended also to take a pro-active stance with survivors, referring some to me for opinion or treatment, and attending legal proceedings in London with victims. In the face of the varied response being provided by social services, the key for me was to find one central contact person in each department, rather than have to deal with area offices and unfamiliar people.

I quickly learnt how we used certain words differently. 'Supervision', to me, meant a dialogue between a disaster worker and someone experienced but less directly involved, with the aim of providing insight about the on-going counselling or therapy, so contributing to the work. For some social workers but not all, it meant a discussion of practical issues with their manager. 'Psychotherapy' was another difficult term, sometimes used in a general sense incorporating counselling and debriefing, but on other occasions used to describe exclusively classical analytical psychotherapy which is inappropriate in the early weeks after disaster. From my medical background, 'patient', 'treatment' and 'medication' are firmly ingrained in my vocabulary, although not always helpful in psychotherapy. They can carry connotations for non-medics of hierarchy and control. 'Research' was another word which caused some difficulty. I saw research in this context as something constructive which would help improve future disaster services. Others saw it as a cynical attempt at self-improvement and damaging to survivors. It soon became clear that those familiar with each other's work could make the translations easily, but for the uninitiated it was not so, and I was finding it as difficult as anyone. All of us had to learn new terms such as 'pro-active' and 'outreach'. Fortunately I discovered that 'mutual support', something I felt a great need for, meant the same in both languages.

Disaster workers

Underneath all this we were all disaster workers with the problems that are common to this work. Often the work was performed without extra pay and in an atmosphere of some management's seeming disinterest. We all suffered the conflict between this and our other work, also with trauma of various types. It was difficult just trying to provide Clapham victims with an adequate service. We all had to learn a lot about the accident from those who were there. I often wished that I had seen the British Transport Police video of the rescue early on. Above all we had to listen and plan for long sessions with the survivors and bereaved. The intensity of feeling could provoke visceral reactions. Vomiting at the accident site is described in the disasters literature, and there were times when I found I had made sure that I knew where the waste bin was before hearing more. This was something simply to share with other disaster workers, not with the victim who was painfully recounting to me this same feeling in himself. The reactions can be difficult for helpers to deal with but it is important to feel some of the victim experience before being able to help.

The huge publicity of such an event is overwhelming for all involved. Sometimes it is as if the most private of emotions are out in the open. The memorial service at Winchester Cathedral was very big and attended by politicians and royalty. Some felt comforted and others felt marginalised by such a ceremony. In mass disaster the event becomes publicly owned, and this leaves little room for acknowledging personal grief. Photographers, for example, appeared to seek out the most obvious distress for their photos. The helpers can also feel left out in the open. They are called to account. Criticism is conducted in public, often unfairly, by other disaster workers. At one point I realised that a number of disaster workers whom I looked to for guidance had been criticised publicly. I should have realised that it was only a matter of time before something was written about me – but it still felt unjust when it occurred. In disasters there is a strong and urgent belief in 'the right way' and 'the wrong way', and this seems to be evoked more here than in other work. People quickly take sides about who is to blame. The primitive psychological mechanism of splitting the 'good' from the 'bad' is invoked. Uncertainties seem more difficult to live with, and there is a natural attempt to ward off future calamity with as much knowledge as possible. Destructive criticism adds to the divisions between helpers – something, I suspect, which has involved me at times.

Whatever your background, disaster work is usually a new way of working. Personalities are as important as professional background. Acknowledging this can provide a useful beginning for learning about the work for all those who might become involved. Those who felt most comfortable within Clapham Support were those with whom I had worked in the past, or who had worked within hospitals before. They did not interpret my involvement as 'overmedicalising' the problem. Pre-disaster dialogue between organisations, together with planning and training are perhaps obvious but nevertheless vital recommendations. An interdisciplinary approach is necessary, and rehearsal of the phase two response[2] as part of the local major incident plan is important. This provides an opportunity for rescue staff and hospital staff of all types to see the importance of the emotional aspects of the work, and to be more responsive to the problems in casualties and in themselves.

There were many lessons to be learnt from other disaster workers. People from other services gave me time during the early weeks. The comments I remember were usually from the ends of conversations: 'Don't forget you will need to look after yourself' . . . 'I found that everything I had already learnt in other settings held good' . . . ' You will meet with intense anger and envy – and that is just from other disaster workers'. It is all true but it is difficult to grasp except with hindsight.

The role of the psychotherapist

Does a medic have a role in the phase two disaster response together with the social worker? The answer is yes, for a number of very good reasons. GPs are often the first to be contacted by those who have been injured, whether severely or mildly. Referrals from GPs will tend to be into the medical system with which they are familiar.[3] This was the case after the Clapham accident. Furthermore, for those treated in hospital (a very vulnerable group for long-term psychological problems), arrangements are already in place for appropriate patients to be seen by a member of the mental health team, usually a psychiatrist.

Medics are familiar with diagnosis, injury and treatment, and this is sometimes an advantage in understanding the repercussions of an injury; for example, when there is a head injury that is affecting the mental state. In addition, psychiatric reports are often requested for compensation claims. It is my view that these are usually best provided by someone entirely separate from the ongoing counselling or therapy. Most requests I passed on to an experienced colleague, though a few I did myself after the survivors had completed their contact with me because it seemed unnecessary for yet another traumatic interview to occur.

How large a part should the psychotherapist have? How alien is this profession to people in general? Will social workers, psychologists and psychiatrists engage their help? The answers are as diverse as disasters themselves. As far as the Clapham rail accident is concerned, over 90 people in two and a half years were in touch in some way with Clapham Support, which was based in the psychotherapy department where I worked. Social workers did indeed ask my opinion about certain problems, and one social services department, which was initiating an outreach programme, asked me to talk at a meeting for bereaved and survivors.

In the years since the accident I have been asked to conduct disaster training for social services, voluntary groups and the clergy. Disorders of bereavement and stress following trauma are a major part of psychotherapeutic work with individuals. Much of psychotherapy is helping people to face the distressing experiences of their individual disaster, frequently loss of some type. Most psychotherapists are trained not only in individual work but also group work, which is often chosen by survivors as most helpful to them. Finally, psychotherapists are also frequently already involved in staff support work of some type which, as has been made clear by most of the other writers in this volume, is of central importance to disaster work. It is folly for one person to try to provide all the different facets of care required for survivors and bereaved. Those that try, for whatever reason, will be at best burnt out, and at worst both burnt out and ineffective.

About a year after the accident I applied for funding to research the psychological effects of the accident and the help received by victims. In the course of this I was told that I was likely to be successful because disaster was a sexy subject. I felt horrified by the juxtaposition of such pain and death with the marketability of sex. Thinking on it further though, I knew that the disaster had had an impact on some people's sexual functioning, and led others to take a closer look at their sexual identity and relationships. To this extent the subject does have something to do with sex. On a more lighthearted note, and thinking about my experiences of liaising with other agencies I wondered whether a car sticker should say 'Disaster workers do it on their own in the dark'? Or, alternatively, 'Disaster workers are enlightened and do it together'? The latter may not read so well, but it's much more productive.

Notes

1 The importance of such information is confirmed by Charles Pugh in chapter 6.
2 For an explanation of the nature of a 'phase two' response see: Turner, S. W., Thompson J. A. and Rosser, R. M. (1989) 'The Kings Cross Fire: Planning a "Phase two" response'. *Disaster Management*, 2, 31–7.
3 This is confirmed in Margaret Mitchell's description of the aftermath of Lockerbie in the previous chapter.

9 Responding to the needs of young people after Hungerford

Elizabeth Capewell

> *Elizabeth Capewell is a trainer and consultant in crisis management, and has a background in education, management and community development. Following her work in the Hungerford, Lockerbie and Hillsborough Disasters she set up the Centre for Crisis Management and Education working in the UK and abroad. She was a 1992 Churchill Fellow and has travelled in Israel, Australasia and the United States to develop her work. This chapter considers the impact of a disaster on children and young people. Focusing on youth work, it also looks at the organisational difficulties that those involved in such work are often forced to confront and makes recommendations for those likely to be faced by such work in the future.*

The event

Michael Ryan was a 27-year-old single man who grew up and lived in Southview, Hungerford. He had always been a loner who had difficulty forming relationships and still lived with his mother who had been widowed two years before. He had not been successful in holding down a job, though local community programme workers had managed to find him work, and as part of the programme's publicity he had his photograph taken – the photo which later went round the world's press in very different circumstances.

The killing began when Ryan came across a woman having a picnic with her two young children, aged four and two, in a forest to the west of Hungerford. Having shot her in the back, he left the children, and drove back towards Hungerford. On the way he shot at, but did not injure, the cashier of a service station at which he had stopped. On his return home he began shooting at neighbours, killing some, injuring others. He killed his dog, doused his house in petrol and set it and the rest of the terrace on fire. Ryan shot dead his mother and an unarmed policeman, and killed and injured many others in the town. In all, he fired 133 shots, plunging the community into fear and uncertainty. The police and emergency services had to retreat, telephone lines were jammed, and many children on their school holidays were separated from the safety of home and parents. The dead and injured were dealt with as well as possible, but some could not be reached by emergency services for several hours.

By 2.30 pm Ryan had killed 15 people and injured as many. The last victim was to die two days later, bringing the number of fatalities to 16. As the Youth Leader looked out of the window of the Hungerford Youth Centre in the secondary school that afternoon he saw a figure coming down the drive with a gun slung over his shoulder. Fortunately the children were not in the school or in the adjacent Centre. Eventually the Leader decided to make a break for it, ran out to his car and drove away. Ryan stayed in the school and just before 7.00 pm that evening shot and killed himself.

The significance for those of us in Education was threefold. First, we were, as members of the community, part of the events that took place. Secondly, children and young people were among those who had been shot at, witnessed the shooting of friends and relatives, or just experienced the sense of siege and terror. Thirdly, the schools themselves were right at the centre of the traumatic events. All the children returning to primary school could see the site of the burnt-out houses from their classrooms. Several people who had been injured worked at the school. Those returning to the secondary school did so in the knowledge that an ex-pupil had committed mass-murder and had finally shot himself in a classroom they would have to use.

The Youth and Community Service's involvement

My disaster story begins on a hot summer's day in August, Wednesday 19th, 1987, in Newbury, Berkshire. I was at the time employed as a Youth and Community Officer in the Local Education Authority. I was responsible for the overall management and development of services in West Berkshire, including four of the school-based youth centres. One of these was the Hungerford Centre based in the local secondary school. On that day I had been to the office in the morning but had cancelled my plans to visit

Hungerford. Our house had been burgled the previous day and I wanted to check that my three children were all right, as they had been frightened by the experience. If it had not been for our burglar, I would have been driving over Hungerford Common at about the time that Ryan began his killing spree.

The first indication I received that something awful was happening in the area came about 2.00 pm. Sarah, aged ten, came running into the house with a message from her mother, who worked in a security firm, to stay indoors as there was a gunman at large. My children, already nervous because of the burglary became more agitated and worried that it might be our burglar – perhaps an escaped prisoner who was armed and on the run. By mid-afternoon we heard with great disbelief that the gunman was eight miles away in Hungerford – the last place anyone would have expected such violence to occur. At first, it did not cross my mind that anyone I knew would be involved or that it would be any concern of mine. I was wrong – by the end of the afternoon I received a phone call from the Youth Leader to tell me of his involvement. Apart from being as shocked as the rest of the community, and anxious to know the names of victims and survivors, he insisted he was fine and thankful that he was about to close the Centre and go on holiday for two weeks. I promised to visit him the next day.

I woke the next day full of apprehension, the full extent of the carnage still unknown. Although the UK had experienced several major disasters in the preceding years, little of the experience had been written up, and certainly not for the ordinary worker and manager. Emergency plans were only for the emergency services and for nuclear attack. But I did know that as a manager I had a duty towards my staff and especially the young people we served, and that I needed to go to the Centre to support them and to assess needs. My feelings were complex on the journey next day to Hungerford – disbelief, confusion, fear of the unknown. What will it be like? What role do I have? One couldn't help noticing the excitement of being part of something big and different from the normal wranglings of Youth Club committees – excitement quickly covered up by guilt at feeling such a thing. Hungerford was silent. No-one was about, nothing stirred. The broken window in the classroom where Michael Ryan had shot himself was the only evidence that something had happened. It was a relief when a policeman appeared to check my identity.

I spent a long session with the Youth Leader and another member of staff – all I could do was listen and listen to their stories and, when they had dried up, listen to their stories of other traumatic events and other experiences of death that this had triggered. At the end I tentatively suggested that we needed to offer something to the young people involved. Naturally enough, the Youth Leader was not keen as he was about to go on holiday and the Centre was closing. At this point reality intruded with force – we were asked to leave quickly by the police as there were fears that unexploded grenades might still be in the building. The reality that a violent death had occurred a few floors above, the day before, hit home.

During Thursday afternoon I began to wonder if anything else was being done by any service. In my previous authority I was used to much stronger working links between different departments and agencies. The inter-agency networks in this district were harder to create and maintain. It was through the national news that I heard that the Social Services Department were setting up a Unit to help the survivors and the bereaved. As more information about the shootings came in, it became clear to me that I had a responsibility to investigate the role that our service could play, so I phoned the Director of Social Services and was, consequently, invited to become involved. A phone call from one of the school principals about the anxieties of parents and teachers convinced me of our duty to respond.

Through the weekend and the following Monday I attended a number of planning meetings and began to work out how best to respond. I needed staff who I could be sure had sufficient skills and experience, who were available if needed and who wouldn't be offended if they weren't used, who had empathy with the situation and yet who were not over-involved. I chose to use both local voluntary staff and professional staff from within the County. The Centre could not be run without the local people who knew the systems and young people, but I felt that there also needed to be at least one full-time professional worker present at all times. No-one could guess how young people would be reacting. Trying to contact staff was very frustrating and time-consuming: lists were often incomplete and we had no idea which volunteers were away on holiday.

My two immediate line managers were on holiday but I phoned Shire Hall and reported my plans. They were well received and I was told that if there was anything that I wanted I should let them know. Within a day of this offer the reality of how distant senior managers in the service were from what had happened in Hungerford began to dawn. My requests for some administrative support and a telephone in my office (ordered nearly a year before) were met with excuses. Eventually, I decided that it was easier to try to organise things for myself and managed to persuade a voluntary member of the local District Youth Committee to come in and provide some extra cover so I could be free to leave the office.

My next stop was the Family Help Unit at the Town Hall. It was based in the Upper Hall and I approached with apprehension. Although my invitation to help had been accepted, my role was unclear and I felt the reservations of an outsider intruding on someone else's grief. Inside everyone looked busy and as if they knew what they were doing. I found the unit co-ordinator, a calm and capable man. I explained my plans and what I could offer. He asked me to draw a poster so that the information could be passed around to others. Once my role was accepted and I had something concrete to do, I relaxed and found a quiet corner to get on with the work.

The press officer decided that my poster was good enough to be a press release and he asked me to attend that afternoon's press conference. I was frightened of the idea but guessed that the local press wouldn't be too bad. When I arrived, the world's press was present, and I sat down next to the press officer and did all the relaxation exercises I could think of. It was only then that I noticed the cameras and realised that I was going to be on TV too – and I hadn't even got a comb! I waited in terror as the main speaker was interviewed. Then my turn. I was shocked by some of the questions, but found some inner resource which helped me to talk about my observations and ignore some of the more sensational questions asked. At least the re-opening of the Centre was advertised, though the media insisted on describing the sessions as 'counselling', in spite of specific requests not to mention that word – a complete turn-off to most young people.

The next day I was relieved to get some support from my volunteer clerical assistant. At last I had someone to talk to about all that had happened, my ideas, and the frustrations of organising an effective response. It was wonderful to have someone who could also experience and understand the myriad of enquiries that had to be dealt with and decisions to be made, as well as the bizarre phone calls from very needy people who are often attracted to the scenes of tragedy. She too was struck by the difference in energy and urgency of people who understood what was happening and those who did not. An example of this came from HQ in Reading. My request for clerical help had been denied as they were far too busy to spare anyone even for a few hours. When they eventually responded it was too late. On the other hand, an elected member of the County Council did phone and spoke with a compassion and humanity that showed great understanding of the situation. Even more helpful, he took action immediately on hearing of my need for a phone. It arrived the next day.

As the opening of the Centre approached, more and more members of the press arrived and eventually surrounded the Centre. Needless to say young people could not penetrate such a barrier, but a good number of voluntary youth workers did, and we used the time to share experiences and talk. The meeting gave me the first insight into the wide, unpredictable and complex range of emotions being experienced. It also provided the opportunity to explain why I had felt the need to bring in outside help. At the end of the day I went to bed feeling physically and emotionally exhausted, and thankful that my children were all away on an activity holiday; I noticed, however, my desperate need to keep checking that they were alright.

Gradually attendance at the Centre increased as young people heard that it was open. They valued being able to escape the pressures at home and to be in a place that they felt was theirs. They could take part in any of the activities available, relax, chat, exchange stories. For most young people that was far more acceptable and helpful than being counselled by unknown people in an environment they did not understand.

By the end of the week I felt very overloaded with emotions. I had listened to a multitude of stories, heavily charged with intense feelings, worked extremely hard, made many decisions with little information, met and built relationships with many new people, had rapidly learnt and experienced new skills – all under the probing eye of the world's media. I felt I had done a professional and competent job and had kept my emotions to myself. I had a variety of concerns, however, about having taken charge and control and how colleagues who had been away on holiday would react to what they would find on their return. The sense of urgency for those immersed in disaster work for several weeks contrasts sharply with that of those who have not been involved at all – especially if they have been away on holiday as well. It is hard for the disaster workers to detach themselves sufficiently from the work in order to communicate needs clearly to others in the 'non-crisis world'. One expects them somehow to know, maybe because to you personally it seems so clear. It thus took some time to persuade my manager on his return from holiday that I needed to meet him as soon as possible to inform him of what had happened, and gain support for our work.

In response to the school's requests for help in preparing for the return to school, I had organised two teams of experienced youth counsellors from well-established agencies within an hour's drive of Hungerford. The leaders of each team were to meet with me that day before attending the school staff meeting. My manager arrived at the Youth Centre as I began the briefing meeting. To be thrust into this on his return from holiday must have been difficult and it was perhaps not surprising that some of his contributions showed a lack of appreciation of his role and the needs of his staff and young people.

Once again the contrast in the perspective of people coming from very different places was highlighted and became a source of friction later because the position of neither was stated and heard clearly. On that Monday morning my team were fully into the swing of work which was the result of the hard and intense period of planning and co-ordination of the first week. As much as the support of managers was wanted, it became clear that the perceptions of needs were so out of tune that all we could do was get on with the work we had planned.

The job in hand

This promised to be tough, but we had no idea how tough as we moved over to the school hall to join the first full school staff meeting, the day before the new term started. The complexity and intensity of the dynamics of that meeting remain firmly with me. It reflected the range of emotions I had already observed in the community: grief, anger, sadness, denial, coldness, shock, control, anxiety, euphoria, blaming, support. Our team had been invited in by the Head to help him prepare staff for the return of the children. I concluded that our main role that day had been the 'outsiders' and, as such, the focus of incredible anger by a dominant minority in the staff group. In contrast many teachers privately told us how pleased they were that we were there.

After the Head filled in the details of what had happened and the decisions that had been taken (especially vital for those from outside Hungerford and those who had been on holiday out of the country), he explained what he had asked my team to do. He separated the staff into year groups, and two members of my team worked with each group. We were to listen to teachers' anxieties and help work out various strategies for dealing with different types of reaction. We were particularly concerned about children who would be based in the block where Ryan had shot himself. Those children directly affected by the shootings would be under the care of the Education Psychology Service, but we needed to give attention to the many others who were involved on the day and to those who were away at the time. We tried to help teachers to see how they could manage the students' emotions so that they did not spring up unexpectedly at inappropriate times or develop into hysteria. We struggled to convince them that they had the skills and confidence to do this. Many teachers were still in a state of shock themselves and I believe that much more attention needed to be given to their feelings before they could undertake the management of their pupil's emotions.

My team were left feeling very despondent and worried about the students. We knew, however, that at certain times the only thing to do is withdraw, though we agreed that a few of us should be present in the staff room on the first day of term. The school staff had decided that nothing official would be said in school about the shootings – there would be no memorial as no-one from the school had been killed. On the first day back at school, the only recognition of what had happened in the holidays came from the usual media presence waiting to hustle stories from the children. There were few direct requests for help from the staff, but indirect approaches were more frequent. It became apparent that some felt unable to express their needs because others (by no means all) insisted that it was not necessary and that needing help was a sign of weakness. We made links with individual teachers and did a lot of work by telephone or out of school.

A great deal was also being said about the wonderful pastoral care system in the school, the special nature of the community which made outside help irrelevant, and the resilience of the pupils which meant they would not be affected either at all or for long. I have heard exactly the same comments from other schools affected by a variety of other tragedies in this country and in other parts of the world. This phenomenon has been so marked that it has to signify a much deeper need that has to be met before effective help can be offered to communities such as schools. Disasters attack individuals and organisations at levels which go far beyond logic.

Management issues

This marked the end of my direct involvement in Hungerford. I now had to begin to withdraw and take on a more managerial role assessing long-term needs and liaising with other agencies. However, it also marked the start of my own difficulties which had less to do with the shootings and more to do with the way in which my managers reacted to my involvement in the aftermath of the shootings. As workers were involved in the events themselves and during the aftermath in helping the community, we could not understand why the managers with greatest responsibility for the staff had not phoned either to check on our safety, or to find out from us directly what we had done in the name of the service. The first brief enquiry came some weeks after the event as an afterthought.

It was interesting to observe that managers who valued their pragmatism were not capitalising on the good work done by staff at a time when the Youth Service was under scrutiny. It was also puzzling that no-one wanted to talk in detail about our experiences in order to learn from them, and use the information for the future, either in the County or elsewhere. As well as differing perceptions about the needs of young people and the community as a whole, the way staff working with the aftermath were perceived within Hungerford contrasted greatly with how they were perceived by the department. As the manager in the middle of this tension, difficulties were bound to arise. My colleagues and I had, we felt, done a very competent job in stressful circumstances at a time when little was known about how to respond to such events. We had done this with little involvement from senior managers and assumed a great degree of responsibility. Moreover, this work was high profile and received a great deal of media attention. We had all responded beyond our normal call of duty in a variety of ways using the best of youth and community work skills, and discovering resources within ourselves which would not be tapped in normal, routine work.

In retrospect, my feeling is that the initial inappropriate and inadequate responses from senior managers set up a chain of reactions which led us to respond independently to the needs in the community which we could not ignore. Our anger at this lack of management concern further alienated us from senior staff, which resulted in many months of antagonism, difficulty and misunderstanding. For people so involved in this intense work, it is hard to step back and see that it may be one's own anger that prevents people from hearing and responding appropriately. Much of this would most likely not have happened if time had been given early on for all staff to tell their stories and be heard. I had found this an invaluable exercise with the Youth Leader and the voluntary staff.

Two years after the shootings I chose to resign my job, not knowing what lay ahead, but determined to find answers to the many questions raised for me about disaster management and the needs of young people. Once I had resigned, I discovered, and am still discovering, other workers involved at Hungerford who were mismanaged at work and deeply affected by the work. Some retired on health grounds, some are still working but are still affected, some are still angry, one is in prison, and another committed suicide.

What I have learnt is the need for organisations to be wise before the event and do what they can in the way of training, emergency planning, or establishing crisis response teams. Where staff become involved in crisis work, there is a need for managers to become informed of their needs and how they are best managed during and after their involvement. They need to understand the importance of techniques and processes such as psychological debriefing[1] which formalises and gives structure to the need for staff to have their stories heard and their reactions understood. They also need recognition for their work and skills through regular contact from managers – not just to give orders but also to listen and learn from the staff who are close to the needs of the people in crisis.

The impact on children and young people

If these are the reactions of adults who were working with disaster, what of the reactions of children and young people to the shootings. Our own experience taught us about the vital need for people to have the space and permission early on after the event to tell their stories and express their thoughts, feelings and reactions in the safe knowledge that they will not be judged. People need to be helped to identify and develop their own ways of coping, and re-frame their beliefs and thinking in the light of the new knowledge they have about themselves and the world.

In normal times, children and young people are rarely given this opportunity. In times of great tragedy, the general reaction of adults is to shield and protect them from what they perceive as a stressful situation, and prevent them from talking about their experiences – especially in places such as schools. This phenomenon has been observed in all the UK disasters in which children have been affected and has been reported widely in other countries. Equally widespread is the rejection and marginalisation of the professionals who try to recognise the trauma suffered by children.

The common reactions that were reported included anxiety, regression into bed-wetting, clinging to parents and comfort-objects, sleep disturbances, avoidance, fears related to noises, sights and smells related to the shootings. Others suffered because they weren't allowed to talk about what happened or because their fathers, at work and out of Hungerford at the time, did not appreciate what their families had been through. Younger children were able to work through their feelings by playing them out – where they were allowed to do so. Adults found these games painful to watch, but they were very healing for children. However, some now in their late teens, look back and are still puzzled by the way the subject was avoided both at home and in the community. They talk of friends who were affected in some way or other for several years after and had no-one to turn to for help. Even children very directly affected by the shootings did not always receive help that seemed appropriate for them. Even now, one child will tell people that his father is working abroad rather than tell the truth of his death and risk again not being believed by other children.

Another indication of the need of children to express their feelings is shown by a story told recently by a girl who remembers being shaken by the severity of the punishment given to any pupil heard making a joke about the shootings. Joking is a common if unfortunate behaviour following disaster, and is an oft-quoted coping strategy used by the police and other emergency services to defuse the intensity of their feelings. It is possible that for some children this was the only way their feelings could be expressed. Would it have happened if the more structured interventions now available for use in schools had been used to allow controlled and safe discussion of events?

Several examples of work undertaken by the Youth Centre illustrate the value of an educational response where survivors are enabled to mobilise their own resources rather than treating them as patients or cases. The first involves a child who was becoming a real problem for the adults around him. We built up trust through normal youth activities such as a game of darts or pool and general non-threatening chat. Eventually a story unfolded about how someone had been shot outside his house and lain there for hours. That evening the family had gone ahead with their holiday abroad and any mention of the shootings was banned. As with many other children, this young person could not believe that Ryan was dead. With his agreement, he was eventually ready to be taken up to the room where Ryan had killed himself, talking through his fears as he went. Once listened to, and once his questions were satisfactorily answered in a familiar environment, he was no longer a 'problem'. His mother was amazed at the change in him.

The second example shows how an outlet can be provided for all the young people in the area affected by the disaster. They may be experiencing emotions they don't understand because they were 'near-misses' or because they identify in some way with the survivors and community. Offering help is a positive and useful way for young people to respond to a disaster, just as it is for adults. However, offers of help from young people are not often received well by those organisations that are involved in the aftermaths of disasters. At the instigation of local youth workers, a fund-raising 'Paddle for Hungerford' in canoes on the Kennet-Avon Canal gave over two hundred young people the opportunity to express their sympathy and do something useful in their own way.

The third example relates to the public rituals which help communities acknowledge an event and then move on. When whole communities and the area around are involved in mass grieving, even the adult survivors feel that their wishes and needs become submerged. In the public memorials young people especially can feel excluded. In the year following the shootings this was gradually addressed at the Hungerford Centre. Young people themselves decided that this important event in their lives needed to be acknowledged, but in a way that expressed hope for the future. They decided to provide a living memorial to the life of Sandra Hill who had been shot dead on her way to visit the Youth Centre where she had been an active senior member. A Youth Award was established to be presented each year to the member who had given most to the management of the Centre. The Award was presented for the first time eighteen months after the shootings, in the presence of Sandra's parents and at a ceremony organised and run by young people. It was a moving and memorable occasion and one which allowed people to talk of things that were difficult to mention elsewhere.

In conclusion, the following are a few essential steps to be taken when helping children and young people in the aftermath of a disaster:

- Assume that children and young people, including babies, will be affected in some way by disaster and that the ripple effect will be much wider than you would ever imagine. Find out who is involved. This step is often missed out.
- Create environments in places familiar to children that are part of their normal experience (home, school, clubs) where it is safe for them to ask questions and express feelings and thoughts. For example, set up small group discussions, art or storytelling sessions – it doesn't matter about the subject, the feelings will come out.
- Do not wait for children to talk or show symptoms. They may not have the language to describe new and strong emotions. Learn to pick up the signals that help is being asked for before they resort to signals that adults don't ignore – like bad behaviour.
- Learn to deal with your own emotions first so you can provide an example for children of how they can deal with theirs. You will also be able to help them form a clearer perspective if your own feelings are acknowledged. All too

easily adults can make children into the victims needing help – it is easier than accepting help for yourself.
- Where possible, help the whole family or class to support each other – they are a delicate system affecting each other.
- Children are a part of the community and the means by which the experience will be integrated into the memory of the community. They therefore must be included in the community rituals and healing processes.
- Allow young children to play and draw their stories and feelings, however much it upsets adults watching. It only becomes a problem if the same things are repeated over and over again for several weeks without the child finding any relief.
- Children less involved need information about how to react to friends who are more affected.
- Much can be done through school to prepare children to deal with stressful life events, which will help them if disaster strikes – and if it doesn't.
- Don't hang on to a survivor who needs medical help. Do what you can but be prepared to consult with or refer to other professionals. Your support is still needed while they are given treatment elsewhere.

Most of these involve attitudes which once addressed free adults to LISTEN to the stories and needs of children and young people. ACCEPT them as valid and PROVIDE OPPORTUNITIES which enable them to obtain information, express their feelings, and create rituals and memorials which help them move forward. Nothing that happens to a child (or an adult) on this scale can ever be forgotten, but a place can be found, with help, for the memories so that they are not stopped from living life to the full.

Finally, there are a number of important things that schools can also do:

- Prepare pupils through specialist programmes such as stress prevention, death education programmes, and peer support skills within the normal curriculum.
- Plan for emergencies – have policies and plans in place for a) dealing with bereaved pupils, and b) dealing with emergencies.
- Train *all* teachers to be aware of issues for pupils dealing with trauma, train some teachers to form a crisis response group, and train those with pastoral/counselling roles in work with individuals.

- Post-trauma classroom and Staff Room debriefing and information for staff and pupils.
- Referral-on of pupils needing extra help.
- Create rituals such as memorials and remembering anniversaries to help pupils integrate and move on from the traumatic event.

Notes

1 See chapter 5 by Janet Johnston and Liz Beeson, and Figure 5.1.

10 Reaching out: running a staff care service in the aftermath of disaster

Jane Harper

> Jane Harper *is a trainer and staff care consultant, living in the North-West and working all over the country. She runs a staff care scheme for one local authority for multi-disciplinary staff working in child protection. She has also been involved in supporting staff during court proceedings, inquiries, following violent incidents at work and was, for two years, the co-ordinator of the staff care initiative set up in the wake of the Hillsborough Disaster. Anyone wishing to contact her should phone 061 485 2745. This chapter examines the unique Hillsborough 'Staffline' initiative, the difficulties facing someone trying to support disaster workers, and the staff care lessons learned as a result of this work.*

On a sunny Saturday in April 1989, I was working in my garden when my elderly neighbour told me that 'something awful was happening on television'. I went indoors to watch, stunned, as the Hillsborough Disaster unfolded. As my tears flowed, I was surprised to discover how much they were associated with my beloved brother's sudden death, over twenty years ago. Surprised, particularly, because I had done so much grieving for this long-ago loss during my therapy. However, perhaps not so surprising as he was a fanatical football fan, and, as any survivors of sudden, accidental bereavement will tell you, those feelings of shock, quickly followed by nausea, are only lying dormant and are readily revived when confronted by the tragedy of others.

As a very experienced crisis intervention and grief counsellor, my first thought was to drive to Liverpool to offer my help. However, being sensible, I knew that this affected me too closely and that I might be more of a liability than a help. I was sure that

I could be of more use in a different role. And there it stayed. I am a busy freelance trainer and counsellor and was pre-occupied with a great deal of work at that time. Then, in early May, I received a phone call from the British Association of Social Workers (BASW) asking if I would be interested in being interviewed for a post running a counselling service for social workers doing Hillsborough work. I was indeed interested. I was interviewed and then took up post at the beginning of June. Quite rightly, BASW had wanted someone full-time for the six months this experimental project was due to run. However, we had to compromise, as I could not just abandon my existing clients. We agreed that I would provide cover for three and a half days a week, working from the Northern Office of BASW on Merseyside. The project was named Staffline and consisted of a telephone line, me and some secretarial help from Sheila Gough, the very able person who ran the BASW office. This was the first time ever that professional, confidential help, external to the organisation, had been offered and funded before there was any evidence that anyone needed it; a bold step and a far-sighted move on the part of seven local authorities around Merseyside, with support from their colleagues in Nottinghamshire and Sheffield.

Staffline had been run by a number of volunteer counsellors on loan from other social work agencies for the first few weeks following the disaster. A telephone number – a helpline for staff – had been publicised quite widely, though very few workers used it during those weeks or later. I was certain, from my existing experience of staff care, that it would be difficult for staff to ask for help and that I would need to do a great deal of outreach work, just as the social workers would have to do with their clients. It was likely that the course of my work would mirror the course of the disaster. Furthermore, in reaching out and offering help, I was aware that I faced a credibility gap, just like the social workers. How long would it take to achieve trust and prove that I could be of use? How could one person achieve much in the face of so much possible need? It may be that I would face similar pressures from managers to produce numbers of service users quickly, to justify the risk taken in funding such an ambitious project.

So it was with anticipation, as well as some anxiety, that I began my first day with Staffline. It was greatly supportive to find such a warm welcome and such good organisation from my new secretary. One individual and one team of social workers had already asked for appointments before I had even started, and that was encouraging, especially as I felt I had missed some crucial things by not being involved right from the beginning.

That positive start helped to bolster me for the first difficult event, which arrived before the first day had finished. I attended a meeting of senior managers from the local authorities and other agencies, who had been responsible, from the early days, for the organisation of the social services' response to Hillsborough. Although most managers were welcoming and supportive of the project (though in a rather vague way as they had more complex problems on their minds), I was given a hostile reception from the Assistant Director of one of the participating authorities. He said that they wouldn't be needing me, as they had already set up their own staff support scheme and didn't wish to pay to duplicate the service. I replied by saying that I didn't think this was my problem. I had been engaged by BASW to offer a service to disaster workers and I didn't consider that it was my responsibility to determine who would pay. This was immediately confirmed by other, more senior managers, who reminded everyone that collective agreements had been made and it was expected that they would be honoured. It is interesting to note that this proved to be the only agency where I experienced any political difficulties in the course of my work for Staffline.

So, I had survived my first major obstacle but I had been given warning that six months was going to be a very short time in which to play the 'numbers' game – there would undoubtedly be some managers who would be looking for early proof of demand for my service. I was to find out just how early, some three weeks later, when another manager pressed me about how many clients I was seeing (at that point, formally, three). Again, unbeknown to me at that stage, this was part of a political move by an authority which was perhaps feeling that it had over-committed resources to Hillsborough work, in view of the numbers of actual clients in that district, and was already looking to withdraw from the consortium. This pressure to justify one's existence through the numbers of clients one had was also experienced by the disaster workers, whose teams were often subject to scrutiny about how many people they were counselling, particularly if the politicians in that authority were less than enthusiastic about resources being devoted to on-going work with disaster victims. Again, I was supported by another member of the management group, who pointed out that it was too early to be counting clients. Nevertheless, the pressure felt very real. Looking back, those first few weeks seemed to go by in a blur of travelling, trying to navigate from the A–Z maps whilst driving in unfamiliar territory, meeting lots of new people, attempting to persuade sceptics of the merits of the staff care service. I often think that if I believed passionately in double-glazing, I could make a fortune with all that I learned about selling!

I met with several different responses, depending, I think, largely on the organisational setting of the workers, but also partly on the extent of their involvement in disaster work. Firstly, the most welcoming response came from workers in the large Shire counties, who were carrying Hillsborough work alongside their normal workload and who commonly felt isolated and under-supervised. Often they had been allocated a team leader especially for this work but this person frequently felt equally unsure about how to cope with it. For these team leaders, expressing *their* uncertainty and *their* need for support was usually not too difficult, as they did not consider this to be their 'real' job. Consequently, for staff in those authorities, someone who was interested in how they were coping and who also knew something about the work was a welcome sight indeed.

There was also a warm response from the team that had been established very close to the Hillsborough ground in Sheffield. I think this was, at least in part, because of a sense of isolation from Merseyside where, by now, the bulk of the work was taking place. In addition, as time went on, I began to become aware of a massive sense of avoidance in Sheffield. It was almost as if, collectively, in some strange way, people in the city wanted to forget that this tragic event had happened at *their* football ground. After all, Sheffield were the neutral parties. Their team wasn't even playing in the semi-final of the Cup. So the small group of social workers and their administrative staff carried an enormous burden on behalf of their City. Under difficult circumstances, in which there was considerable pressure to act as if the disaster was 'all in the past', these workers tried to keep alive, to staff at the ground, the importance of offering help from Sheffield to other professionals who might need to talk (police officers for instance); to social workers bringing bereaved people and survivors on harrowing journeys to retrace the last steps of loved ones and see where they died, or to face again the place where they watched people die or felt their own life slipping away as they struggled for breath; and also to provide a local, helpful presence at the long and exhausting inquests held in the city.

Lone workers in parts of the country very distant from Merseyside, Sheffield or Nottingham welcomed my contact and invitation to a series of 'conferences' (of which more anon) where they could meet up with other people doing Hillsborough work. They reacted a little as if the cavalry had just appeared over the hill. It was then that I began to realise just how difficult it was to sustain this work alone and unsupported. The following comment from one of the social workers illustrates what they took to be the value of the 'Burton Manor' conferences:

> The main thing was just meeting up with other people; hearing them say what you're thinking. Because when you're on your own it is easy to think 'I'm the only one doing this' or, 'if anyone heard me say this' . . . I think the Burton Manor days were quite low-key . . . not too structured. [They] gave you a chance to talk to people, to make allies. It was like one long de-briefing system.[1]

This had been exactly what I had intended. I remember, though, having anxious moments about whether participants would think they were unstructured and perhaps a waste of time.

Workers who had been involved in the early days following the disaster, but who may have been no longer in contact with survivors – for example many youth workers who had worked at Anfield during the first week or so after Hillsborough, when thousands of people came to express their sympathy and sense of loss – tended to approach group debriefing sessions (of necessity, late) with understandable caution. Indeed, it was often the case that speaking about that time brought forth many tears and, commonly, a great deal of anger, mainly directed towards managers who 'did not understand'. Despite the upset, most of those workers welcomed the opportunity, retrospectively, to talk about their experiences.

The most guarded, sometimes bordering on hostile, response came from the specialist social work teams on Merseyside. There were times when I was treated as if I were offering nuclear waste rather than support with a very difficult job! Thankfully, I was able to hold onto my sense of humour, aided by the indomitable Sheila as well as by my understanding of the pressures and anxiety that might lead to such behaviour. However, I have to say that there were days when I went home wondering what had induced me to take on such a crazy task.

Hillsborough Disaster work was high profile work, with considerable interest from the media and from local politicians and, at times, resentment or envy from other workers. There was, consequently, a good deal of pressure to 'get it right'. This meant, I think, that outsiders were often treated with suspicion. There was also, at times, a sense of competition among the teams; another factor in people not letting you see what they were up to.

The key players in the lives of the specialist social work teams were undoubtedly the team leaders. They often showed a heightened sense of responsibility for their workers and I got the impression that they thought it would appear that they weren't doing their job properly if one of their staff needed help from an outsider. Consequently, in terms of the barriers faced by a staff care service, team leaders were the key people to work through. They set the norms and expectations for the team and, I think, there was probably greater dependence on them than would be usual in a social work team, because the work was so uncertain and so emotionally draining. The most extreme example of the power of people in such a role to set the tone can be seen in the following quote from a member of one of the specialist teams:

> our team leader told us about someone from another team who had sought help. He said, shouting and banging the table, 'THAT IS NOT GOING TO HAPPEN IN THIS TEAM.'

One can imagine how difficult it might have been for someone in that team to come for help. In addition, because of such views, confidentiality was also a problem. Rumour travelled quickly in the region. Confidentiality was something I had to be scrupulously careful to maintain. One slip and the reputation of Staffline would have been shot to pieces. It is entirely another matter if an individual herself/himself chose to tell others that s/he had been a user of the service.

My first strategy then was to meet as many people as I could in the first few weeks and to try to prove that I could be of practical use by recommending books, putting people in touch with one another, and, sometimes, providing training or consultation sessions directly. The next question was how to take Staffline forward as a staff *support* service and not just a staff *counselling* service. I decided to take advantage of an offer that had been made by Burton Manor College, a residential conference centre on the Wirral, to provide free facilities for a conference for Hillsborough workers in the summer.

This offer had originally been turned down by managers as it was thought to be bad timing, due to staff leave over the summer. Despite the timing not being ideal, I had learned just how isolated many staff were feeling, and yet I had also seen the many creative ideas being used by some teams to engage survivors of the disaster. I thought there was everything to be gained by getting workers together: they could share ideas, put faces to names and provide some much-needed mutual support, even if some staff were unable to attend. Consequently, I thought it was worth the risk to go ahead and advertise the 'conference' nationally.

I guessed that it would be more acceptable to workers and managers to label this as training, even though it was more in the nature of networking and skills sharing for, once again – 'mirroring' issues from clients – many workers expressed a kind of survivor guilt; it didn't feel right for them to have treats and get taken care of, whilst so many clients were suffering. We welcomed 80 workers from all over the country to Burton Manor on a beautiful day in August. Very few, despite travelling some distance, took advantage of the offer of free accommodation, but we did ensure that a wonderful lunch was served outside. The lunches and the warmth of the contacts made there ensured a good attendance at the following three 'conferences' that I arranged at the same venue in subsequent months.

The morning of that first conference was taken up with groups making collages symbolising their own experiences of, and feelings about, the Hillsborough disaster. I felt some trepidation about whether participants would wish to get involved in such a personal exercise and whether it might evoke memories which would prove too painful for such a public occasion. However, the exercise proved to be a wonderful start to the sharing of knowledge and experience and the groups produced very moving work. In the afternoon, we were able to network and share ideas as well as plan for the start of the football season and the work that might bring. These themes of sharing knowledge and forward planning were to appear again in subsequent conferences, particularly one which focused on 'endings'. Being open about feelings and the stresses of the work did much to begin changing the widely shared feeling that it was shameful to need help. It also gave me an opportunity to get to know and be seen by many of the workers, in a relaxed setting. Following each conference, there were always workers waiting to make appointments for counselling. Indeed, at the first conference, there were people who needed to be seen immediately; particularly one worker who had actually been present at the match itself.

So, as we approached the end of the six month project, I had met almost all the key workers who had been, or who were still involved, with Hillsborough work. One successful conference had taken place and another one was in the pipeline. Several groups of workers had had the opportunity to debrief about their experiences in the early days following the disaster and several teams and individuals had sought consultation about aspects of the work that were concerning them, particularly uncertainties about working with children's grief and coping with anxieties about suicide risk.

As far as individual counselling went, I had seen six people formally, one on a long-term basis, and many others had sought information/confirmation about their reactions to their work. One of the specialist teams had asked me to meet with them fortnightly, to take over from their external consultant, who had been helping them since the beginning of the work. I was also in regular contact with the other teams. I had attempted to provide some sort of service for workers in Sheffield and Nottinghamshire, even though they had not been part of my original brief, but time and travelling constraints meant that this service was really only scratching the surface of their needs. Staffline had established itself sufficiently to be extended for another six months. One of the workers who had fairly lengthy contact with me commented that:

> it gave me a clearer perspective about my role in that situation (Hillsborough work), and also about my role more generally, and about me. If counselling is about helping you to see, to grow and develop, and gain a greater acceptance, then in success terms, the time I spent with Jane was very successful . . . A very, very traumatic experience though.

Just as the disaster workers were finding that many people's needs began to surface after several months, so too, I found once I was known and trusted, demand for the service began to grow. Indeed, the project was extended again, lasting two years in all (although in the second year my hours dropped to half-time). By the second year, I was fully booked with counselling sessions on most weeks. Altogether, I saw 21 clients for formal, individual counselling sessions, which were directly connected with helping those affected by the Hillsborough disaster. The work ranged from one-off sessions to long-term help for a year. I don't think it is any coincidence that I eventually saw two-thirds of the staff in the longest running team.

People sought help with problems which fell into three main areas. Some workers had concurrent life issues which felt similar to the losses being faced by clients. For example, they might have been going through a divorce or recently experienced a death in the family. Others found that working day after day with people experiencing great anguish and fear, evoked painful memories from their own childhood such as parental cruelty or a parent's tragic death. These feelings had usually lain dormant for many years. The third big source of stress for many staff was when particular times or events brought about an upsurge in their work which proved very tiring, for example, the first Christmas following the disaster or the lengthy and painful inquest held almost two years afterwards. Many more staff consulted me about anxieties concerning cases, and a considerable amount of work came about as a result of concern about closure of teams, for two reasons: either because there was a good deal of uncertainty about whether teams would be kept going, thus making it impossible to plan work, or because staff felt that the closure of their team was premature or being handled insensitively.

It was very difficult; at times, to separate out what could be regarded as counselling and what might be seen as consultation[2].

Some people came at different times for both. Others, who would probably not have considered themselves as counselling clients, were felt by me to have been offered a good deal of therapeutic help in amongst the consultation about their work. What matters, after all, is not what we call it, but the fact that people received much-needed help and were able to carry on offering a skilled service to others.

As far as the management tier was concerned, I think they found it even more difficult to acknowledge the need for, let alone ask for, personal help. I saw one team leader for several sessions and one middle manager from a distant authority who had been badly affected by being present at the match. Unfortunately, I was too far away to be of consistent help, but was able to suggest a local resource. I did see one senior Youth Service manager for counselling and tried to have lunch sometimes with Social Services managers, so that I could be a listening ear if they needed one. With only one exception, I had unfailing support and backing from senior managers, and I wanted to give something back, where possible.

It remained difficult for workers to look for support and those who did were, of course protected by the anonymity of the service, for as one worker commented:

> it would have been very risky in our team (to admit to using the service) because it would have been quite hard to have said in that team 'I'm struggling'.

However, one of the things that pleased me more than anything else was that almost everyone who came for counselling ended up telling their colleagues, and, in most cases, their managers too. In this way the view that seeking help was a sign of weakness was slowly broken down. It certainly helped to spread the word. The following statement illustrates the importance of the informal networks:

> ... the time when I looked to staff support ... there were other people (in the team) seeing Jane around that time and I can't remember whether it was someone suggesting it or whether it was me deciding ... so I picked up the phone.

During subsequent involvement in a staff care scheme for child protection workers, I have found, once again, that the informal network is by far the most crucial one in determining user levels.

A not unexpected issue for Staffline was the dearth of male clients at first. My experience in other staff care schemes bears out the fact that men seem to find it even more difficult to ask for help than do women. Also, of course, the majority of managers were men and the difficulties for this group has already been discussed. All the people I saw for help in the management tier were men and all said they had found it beneficial. However, they were the group least likely to let others know that they had seen me, so the 'informal grapevine' could not work amongst this group in the same way that it did amongst the social workers. By the second year of the project I had become much more 'visible' and better known to the social workers. Also, the long-term nature of the work was taking its toll. Perhaps only then did it feel acceptable to the male staff to acknowledge their need for help, and many more of them became clients in that year.

Lessons from the staff care service

So what are the key things I learned from my two years running Staffline? Overwhelmingly, that the 'outreach' model is just as important for staff care as it is for disaster work. Also, that it takes time to build contacts and establish trust, so early pressure to produce significant numbers of service users is very unhelpful. A crucial factor is that the service is offered by an organisation, independent of, and external to, the employing organisation. But more than this, it was extremely helpful to have senior management support (on occasions this was used to effect positive organisational change) and to have the luxury of knowing that the service was paid for in advance. In other staff care work I have encountered considerable difficulties at times in persuading managers that traumatic events cannot necessarily be dealt with in two sessions! The other consideration, of course, is that if someone senior has to authorise payment, it is impossible to maintain confidentiality about who is using the service.

I learned that it was possible to make an impact on the 'macho' culture which assumes that it should be possible for workers to cope with all things, at all times, preferably alone, and that needing help, therefore, is a sign of weakness and failure. I also discovered quickly that the role of the team leader was the most crucial one in terms of gaining access to those working with disaster-affected people. This was also the group who were least likely to have had training in disaster work and who were often poorly supported by senior managers.

The aspects of 'mirroring' were, at times, uncanny, and I certainly needed more support myself when there was particular stress for social workers, e.g. during the inquest proceedings. I did have a Steering Group for the project, consisting of a small number of local BASW activists and one senior BASW employee. In the first year of Staffline, they met monthly and this was supportive but they were just not able to be there enough to fulfil my needs, and these meetings grew more infrequent in the second year. Members of the Merseyside branch of BASW were very helpful and my secretary, Sheila, was a tower of strength. She also had unerring skills, in my absence, in assessing who needed to be seen urgently, so I knew the project was in safe hands. Fortunately, I already had my own consultant for my other therapeutic work and was also able to use her to support me with Staffline work. There were still times, however, particularly in the early days, when I felt very stressed and it was hard to find someone to talk to who understood the problems. I was very aware that confidentiality was crucial. One slip would have jeopardised the whole project. On reflection, it would have been better from the support point of view to have had two workers sharing a full-time post, but that might have made trust and credibility harder to attain.

Although the early days felt like wading through treacle, I never got stuck and I can honestly say this was the most interesting work I had ever been involved in. For years, whilst training professionals involved in crisis and bereavement work, I found myself emphasising the need for staff care. Staffline gave me the opportunity to do it and to prove that it worked. It was good to know that it had really made a difference and that we had, by and large, prevented the 'burnout' experienced by staff involved in earlier disaster work. Staffline also gave me the opportunity to meet some of the nicest people of my professional life.

Notes

1 This and all the subsequent quotes are taken from: Newburn, T. (1992) 'Caring for the Carers: The BASW Staffline Project after Hillsborough'. Unpublished report to the British Association of Social Workers.
2 Issues about training are discussed at some length in a report of a research project which I undertook on behalf of the Central Council for Education and Training in Social Work (CCETSW), available from CCETSW, 26 Park Row, Leeds, LS1 5QB. It is important to emphasise here that feeling unprepared to carry out such skilled and exacting work was one of the major sources of stress reported by all workers.

Conclusion – Social welfare after tragedy: what have we learnt?

Tim Newburn

The role of social welfare agencies in the aftermath of disaster only really came under scrutiny in the UK in the 1980s. There is now an assumption that, for example, social workers will become involved in post-disaster 'counselling', that a variety of professional specialists will provide 'treatment' of one sort or another, and that voluntary agencies in the environs of the disaster will be mobilised to provide food, transport, shelter and a number of other services.

One thing that is now clear and widely accepted is that there is much in the way of pre-planning and organisation that can be undertaken. Despite the fact that, almost by definition, disasters are unexpected, much can be done in advance of them to ensure that if and when they do occur, those likely to be tasked with responding to the needs of those affected are as well prepared as possible. This view now has government backing – the Department of Health has sponsored the production of a report that provides the most up-to-date guide to post-disaster organisation, the Home Secretary now has a specially appointed civil emergencies advisor and there is a national emergency planning college which focuses on peacetime as well as wartime emergencies run under the aegis of the Home Office.

Conclusion 131

There now exists within the public, private and voluntary sectors considerable experience of post-disaster work. Some of this experience is shared in the conferences and seminars that now proliferate, in the increasing number of training courses that are available, as well as in some of the new disasters literature. However, there remain large gaps in our knowledge, particularly in relation to the type of work undertaken by social services and related organisations after disaster. It is in response to this gap that this book has been written, and its general aims are twofold. The first has been to draw out certain key lessons about the structure and content of disaster support work and, secondly, to attempt to do so in a way that gives some idea of both the nature and the consequences of undertaking such work. Consequently, the information has been presented in an accessible and easily readable form, avoiding, as far as is possible, academic language and the conventions of academic literature such as footnotes, references and so on. Furthermore, although focusing on social welfare issues, a much broader range of experiences has been reported than, say, just those of social workers and other social services staff. Post-disaster support potentially also involves police officers, general practitioners, lawyers, youth workers, psychiatrists, staff counsellors and, of course, many others.

Given the number and variety of organisations and individuals likely to be involved in such work it is vital therefore that each appreciates and values what the other can offer (and what it will not be able to deliver), and that constructive and strong links between agencies are developed and maintained. In the majority of contributions in this volume there were references to the difficulties associated with the setting up and maintenance of co-operative working relationships with other groups. David Mason's job, for example, consisted in large part of bringing together a number of authorities under one umbrella, and once there, attempting to keep them there. One message that cannot be repeated too often is that in the aftermath of disaster there is frequently, if not always, some element of competition between agencies. This must be anticipated and measures taken early on to try to ensure that the worst elements of this are mitigated.

One of the crucial lessons from Hillsborough was that it was vital that there was only **one** helpline. This minimised confusion and ensured that there was a central point from which referrals could be monitored and staff organised. Although helplines have become perhaps the centrepiece of 'welfare' responses to disasters, it is recognised that this can only be a partial solution to the 'problem' of offering help to those affected. The reason for this is that helplines put the emphasis on the person needing

Martin Hinks describes, at best it may take a long time for people to accept that the way they are feeling may not change unless they use some of the services that are available. This is vividly illustrated in Margaret Mitchell's account of the role of GPs after Lockerbie, where it seems clear that many people in the community would not have received help had they not been forced to consult their GP because of compensation claims. The simple lesson here is that a greater number of people are likely to be reached if services are not merely available, but are *offered*. This in itself requires many professionals to rethink the ways in which they work and to re-examine their skills. *Offering* help – especially in the context of the aftermath of a disaster – requires a sense of security and purpose that need not be present in those tasked with responding to *requests* for help.

The very different character, or at least context, of 'disaster work', has implications for the training of social workers and other professionals. Renewed emphasis needs to be placed on 'listening skills' in professional education, and specific post-qualification 'orientation' training should be provided for any staff who are likely to become involved in this work. Many of the contributors to this volume describe their unpreparedness for this very difficult work, and whilst workers can never be fully prepared for such an eventuality, they can, at least, be provided with as much information about what they are likely to confront as is available. Furthermore, as several of them testify, it is often those who have been through such experiences themselves who are the best source of such knowledge. Such people are still all too infrequently used by training departments and agencies and it is time this situation was remedied.

Given the difficulties both in recognising the need for and in asking for help, it is clear that services must be available in the long-term if they are to be effective. The recent Department of Health recommendations suggest that two years is a reasonable length of time for authorities to anticipate having to keep services in operation, though Martin Hinks' and Elizabeth Capewell's experiences suggest that even this may not be long enough. Nevertheless, if such a recommendation were widely adopted a significant advance on the current situation would have been achieved.

Whatever the circumstances, services can only be used if they are known about. Thus, for example, as soon as plans for setting up a helpline are put into operation, so arrangements must be made for publicising the service. Many of the contributors to this volume make reference to a disaster leaflet. With certain local changes made to personalise it, this leaflet has been identical in each case, and has been widely used in both Australia and the UK to explain to the general public the potential impact of disaster, and to provide details of where help can be found. Entitled 'Coping with a major personal crisis', all agencies should have copies of it that can be quickly reproduced in the event of an emergency in their vicinity[1].

Another ever-present factor in individuals' accounts of their involvement in disaster is the mass media. Much of what is said about the presence of the media in the aftermath of disaster is extremely negative and critical, particularly in relation to the behaviour of some tabloid journalists and photographers. Indeed, David Mason confirms that he came across some significant examples of misbehaviour by journalists after Hillsborough, and Margaret Mitchell suggests that the presence of media after Lockerbie was also, on occasion, upsetting. What comes across strongly in both these accounts, however, is the importance of attempting to establish positive and constructive relations with press, radio and television journalists. The way to achieve this, they suggest, is to take the lead in providing information about the disaster and about services. As David Mason reported, for example, the initial crush of reporters at the headquarters of social services on the morning after the disaster was relieved by the promise of a press briefing at a stated time later in the morning. Similarly Margaret Mitchell describes the 'control' of the press as being one of the primary roles of the police in the immediate aftermath, and that this was also done, in part, by holding regular briefings.

As I have already suggested, the provision of information as quickly and widely as possible is vital after a major tragedy. Consequently, the mass media have a vital role to play, and they can only do this effectively if they are provided with reliable information in a form that is usable. They are also much less likely to behave in inappropriate ways if they know in advance when they will be given information, and as long as they believe that this is as full and as complete as possible. Nevertheless, there will no doubt always be some who will be only too willing to cross the boundaries of acceptable behaviour in search of a good story. Social services and other agencies therefore need to be geared up to provide information in a constructive way. This means that they not only have to take on board the idea that they can work in partnership with press and television, but that they must be adept at producing information, and must have a representative who is experienced in such matters, or has at least been given media training.

More than one of the contributors talked of their fears about the media, and given the demands that the press, radio and television make (and will continue to make) after such events, it is vital that media, and given the demands that the press, radio and television make (and will continue to make) after such events, it is vital that such people are, in future, prepared for such an eventuality. Furthermore, given the level of the demands made, the person given the job of being the main 'spokesperson' for their agency should probably not have any other major duties to perform. At least in the first few days this is likely to be a full-time job. Finally, most large agencies such as social services departments will have a number of press officers on their staff. They will not always be known to senior managers – certainly not to middle managers – yet they are likely to be an invaluable resource under such circumstances.

One of the issues that comes out most clearly in David Whitham's experiences during the aftermath of the Kegworth/ M1 aircrash is the importance of paying attention to the administrative systems that underpin any of the counselling and other support work undertaken. He also gives the example of the telephone helpline. In the first day or two after a disaster, helplines take a huge number of calls, some requesting help, many offering help, and the majority requesting information. As a consequence of this, there needs to be a simple system for logging the calls, for distinguishing their purpose and, particularly where the person is enquiring about someone who has been hurt or killed, some process for providing necessary information whilst ensuring that some element of confidentiality is maintained. Information needs to be checked and double checked, and the police and other emergency services must be kept appraised about what information is being passed out. Administrative work is frequently, if not usually, accorded a very low status by other professionals. However, if the welfare response to a disaster is to be run efficiently it will need to be based upon a coherent administrative system. Indeed, as David Whitham suggests, 'it is worth taking time to set this system up before getting the social work task under way'.

Several of the contributors describe situations in which they found themselves overworking and seemingly unable to stop. This is a common experience in the aftermath of disaster. The early days and even weeks are a time of considerable pressure and strain. Enormous demands are made upon workers both emotionally and physically in terms of the numbers of hours they work. In the immediate aftermath it would appear to be a combination of the amount of work to be done together with a strong sense of urgency that is most likely to undermine staff. Under such circumstances it is particularly important that there is someone who can take control and make sure that some order is maintained – that workers take breaks – for it is clear that the pressure to work combined with the desire to help will make some continue far beyond the bounds of what is reasonable or, indeed, professional. The reality is that there quickly comes a point where workers cease to function effectively and far from being a support to those in need, may themselves become casualties and a further burden on those they are supposed to be helping. Whilst many are able to recognise when the time to stop has come, this will not be true of all, and the expectation that individuals will take care of themselves in this respect is unrealistic. One of the most important managerial functions in the immediate aftermath is to stop staff working at specified intervals and make sure that they go home and do not return until a reasonable period has passed. If there is a lack of clarity about the nature of the organisational response, or about line structures and responsibilities, then an unreasonable onus will be put upon individual workers. This remains true for all organisations working in the aftermath of disasters.

This leads on to an absolutely crucial issue for all those involved in 'disaster work': staff care. This is something which is all too frequently treated as if it were either already sufficiently provided for or, alternatively, merely a peripheral issue of less importance than 'getting on with the work'. It should be clear from the majority of contributions in this book that the current state of affairs is far from satisfactory and is one that must be tackled with the utmost urgency.

The Stafflinc initiative described by Jane Harper in chapter ten had a number of functions, some of which were manifest, others latent. Most explicitly, the primary purpose of the initiative was to support and, on occasion, counsel those workers tasked with providing care in the aftermath of Hillsborough. How this work was to be done, what forms it would take and which workers it would focus on was left largely to the initiative of the co-ordinator. There was sufficient flexibility in the terms of the project to allow for a series of activities that certainly had not been anticipated – for example, the Burton Manor conferences – to take place.

There was, at least at first, considerable mistrust amongst some of the social work staff of the initiative. In general this took two forms. Firstly, mistrust of its primary employee: the co-ordinator. What could she offer? What did she know about disasters/bereavement/social work, etc? Could she be trusted? Secondly, and perhaps more surprisingly there also seems to have been quite widespread mistrust of 'counselling' and support for staff. There were widespread and deeply held views about the need for social workers to remain 'strong', to 'be able to cope' or, rather more extremely, to avoid being seen as 'someone who is likely to crack up'. Part of the work of the co-ordinator of Staffline – indeed a large part – centred around a variety of activities aimed at establishing credibility, not only personally, but crucially, for the idea that it is reasonable for helpers to ask for help.

The term 'credibility' in this context meant several things and was tested in a number of ways. Most obviously, the 'counsellor' was not, at first, trusted by many of the staff. They doubted the worth of the service and, more particularly, they doubted the abilities of the counsellor. At the heart of this lay the highly stigmatised notion of 'client', which is a label that workers will go a long way to avoid themselves. In order to overcome scepticism and even hostility the staff counsellor had to engage in a variety of more or less practical tasks. The manner in which these were addressed and dealt with determined to some extent how *credible* the counsellor became in the eyes of the workers – the potential 'clientele'. With disaster work at least – although there is no reason to suppose that this cannot be applied more generally to 'specialised work' – there is another aspect to the question of 'credibility'. This can be summed up as 'knowing but not knowing'. It is important for the counsellor to be trusted, to be viewed as someone who understands what the work entails and involves, but who is yet in some fundamental way outside the everyday structures and concerns of the work. This can, understandably, become something of a precarious balancing act. Personal and professional credibility is difficult to achieve at first, but it is essential to the success of the enterprise.

Two elements of the Staffline initiative – the fact that it was extra-departmental and that a considerable amount of outreach work was undertaken – were also central to any successes it achieved. Jane Harper's view was that many of the workers would have been most unlikely to seek help or support from a counsellor who was located *within* their own organisation. Although few workers used the service without their line-manager knowing that they were doing so, most expressed doubts about whether such a service would be used if it were not 'independent'. In the main, they feared that intra-departmental services would, in reality, not be truly confidential, and that using them would adversely affect future career opportunities. Elizabeth Capewell's experiences reported in chapter nine would appear to bear this out. Furthermore, many of the workers who did eventually use the extra-departmental service that was provided after Hillsborough were only engaged because the Staffline counsellor had, in effect, gone out and recruited them. As they themselves admitted, she had to do this not because they weren't in need of support, but that they would not *admit* that this was the case. The clear message here is that the model for a post-disaster staff care service should *not* be the helpline model.

As a consequence of the reluctance of social workers to ask for help or to 'admit' to needing support from an outside agency, one of the important latent functions of such a service is as a vehicle for breaking down resistances to staff counselling and myths about staff support. Shame and stigma about 'needing help' are by no means confined to the more macho professions. Social workers appear to be just as reluctant to admit that they cannot or will not be able to cope with everything that confronts them in their jobs as are policemen or firemen, if not more so. One of the greatest successes of the Staffline initiative, even though it was focused on a small number of workers, was that it encouraged a healthier view of what 'carers' are able to do. By implication it also forced some of them to reflect rather more critically on their views of their 'clients' and what they were providing for them.

A further lesson that has been learnt in recent years is that children are profoundly affected by disasters. Whilst, on the surface, this might seem an unremarkable, even trite message to be conveying, it is only recently that children's 'victimisation' has begun to be taken seriously[2]. Children have been both the direct and indirect victims in many recent disasters. However, one consequence of their lack of 'visibility' is that it has often been all too easy to overlook the impact of tragic events upon them. A couple of examples will suffice. William Yule[3] tells of a case in which a boy whose GCSE coursework had been destroyed in the *Herald* Disaster. When the boy returned to school, quite understandably he found it difficult to concentrate on on-going work and yet was still pressed to complete the missing work; work that was, of course, now associated with the disaster. That forcing him to do the work might be emotionally upsetting was seemingly not considered. Similarly, many social workers involved in responding to the Hillsborough Disaster found that the approaches that they made to schools were occasionally rebuffed with the message that the pupils should not be reminded about the disaster because they had their forthcoming exams to think about.

Elizabeth Capewell also experienced resistance from adults to the idea that formal support should be provided for children and young people. She argues that this 'protectiveness' is counterproductive and that it is important to find ways to allow children to express their feelings and emotions in an environment which is safe. In fact adults tend not to deal with children's reactions to such events in order to avoid having to acknowledge their own feelings about this or other incidents and, as Charles Pugh argues, professionals sometimes adopt precisely this kind of 'protective' attitude towards clients as well. Elizabeth Capewell provides a variety of examples of ways of working with young people, and they are, in the main, little different in approach to the work that would be undertaken with adults: she stresses the importance of outreach; of working in a familiar setting; of using groupwork; of pre-planning and of information – especially for those less directly involved.

The point has been made repeatedly that the provision of information is absolutely central to the 'welfare' task after disaster. Those affected are left with a bewildering variety of questions about what happened (to themselves and/or to their loved ones) and why it happened. There are a variety of ways in which social services and other organisations can and should attempt to extract information that will be of interest and importance to the victims of the disaster from the police and other emergency services, and from other authorities. One might assume that the array of legal proceedings that inevitably follow disasters would be a source of clear information for bereaved relatives and others who were affected. Experience over the last ten years, however, suggests that this is far from being the case. As Charles Pugh argues, Coroners' inquests in particular appear to be a most unsatisfactory forum as far as relatives are concerned. Their questions about where, how, when and why their loved ones died are, for a number of reasons, unlikely to be answered. Perhaps most importantly, the inquests are usually conducted as one hearing. That is, the deaths are treated *en masse*, with evidence of a general nature being presented, rather than the specifics of each individual case being explored separately. This is highly unsatisfactory for the bereaved, and Charles Pugh recommends that individual inquests tailored to these relatives' 'need to know' are the solution and, furthermore, that lawyers have a vital role to play in ensuring that relatives are kept properly informed. As such, they have potentially an extremely important role to play in the 'welfare response' to a disaster, and one that is not well understood. Furthermore, it is one that the majority of lawyers – and this would certainly also apply to GPs – are ill-prepared to take on. These professionals need to be prepared for this role and 'disaster training' or, at least, familiarisation, is something that should be made widely available and not confined to the emergency services and, occasionally social services, as is currently the case.

The consistent message throughout the book is that there exists a great array of skills that can be brought to bear in support of those affected by major tragedies. The sources of these skills need to be identified, and then plans made for their co-ordination and control. There is an urgent need for considerably increased general education and training about the realities of the impact of disaster, about what skills and approaches are needed and effective, about how a much wider range of organisations can become involved co-operatively in such work, together with a greater sense of responsibility on the part of 'welfare organisations' for the welfare of their own staff. For, as the contributions to this book have made clear, undertaking such work is both rewarding and stressful. Professionals must be both sympathetically managed and properly supported if they too are not to become victims of the disaster.

Notes

1 The text of this leaflet is reproduced at the back of the book.
2 Thus, for example, in the area of criminal victimisation, it is only in recent years that support organisations have begun to take on board the importance of considering the needs of children. See Morgan, J. and Zedner, L. (1992) *Child Victims: Crime, Impact and Criminal Justice*. Oxford University Press.
3 Yule, W. (1990) 'The effects of disasters on children'. *Bereavement Care*', 9(1), Spring.

Appendix: (Standard disaster leaflet) Coping with a major personal crisis

Somebody you know may have died or been injured. Your experience was a very personal one but this pamphlet will help you to know how others have reacted in similar situations. It will also show how you can help normal healing to occur and to avoid some pitfalls.

Normal feelings and emotions always experienced

Fear
— of damage to oneself and those we love.
— of being left alone, of having to leave loved ones.
— of 'breaking down' or 'losing control'.
— of a similar event happening again.

Helplessness
— crises show up human powerlessness, as well as strength

Sadness
— for deaths, injuries and losses of every kind.

Longing
— for all that has gone.

Guilt
— for being better off than others, i.e. being alive, not injured, having things.
regrets for things not done.

Shame
— for having been exposed as helpless, 'emotional' and needing others.
— for not having reacted as one would have wished.

Anger	— at what has happened, at whoever caused it or allowed it to happen.
— at the injustice and senselessness of it all.
— at the shame and indignities.
— at the lack of proper understanding by others, the inefficiencies.
— WHY ME?
Memories	— of feelings, of loss or of love for other people in your life who have been injured or died.
Let down	— disappointments, which alternate with
Hope	— for the future, for better times.

Everyone has these feelings. The experience of other disasters has shown that they may be particularly intense if

— many people died
— their deaths were sudden, violent, or occurred in horrifying circumstances.
— there was great dependence on the person who died
— the relationship with the person was at a difficult stage
— this stress came on top of others

Nature heals through allowing these feelings to come out. This will not lead to loss of control of the mind, but stopping these feelings may lead to nervous and physical problems. Crying gives relief.

Physical and mental sensations

You may feel bodily sensations with or without the feelings described. Sometimes they are due to the crisis, even if they develop many months after the event.

Some common sensations are tiredness, sleeplessness, bad dreams, fuzziness of the mind including loss of memory and concentration, dizziness, palpitations, shakes, difficulty in breathing, choking in the throat and chest, nausea, diarrhoea, muscular tension which may lead to pain, e.g. headaches, neck and backaches, dragging in the womb, menstrual disorders, change in sexual interest.

Family and social relationships

New friendships and group bonds may come into being. On the other hand, strains in relationships may appear. The good feelings in giving and receiving may be replaced by conflict. You may feel that too little or the wrong things are offered, or that you cannot give as much as is expected. Accidents are more frequent after severe stresses. Alcohol and drug intake may increase due to the extra tensions.

The following make the events and the feelings about them easier to bear

Numbness — Your mind may allow the misfortune to be felt only slowly. At first you may feel numb. The event may seem unreal, like a dream, something that has not really happened. People often see this wrongly either as 'being strong', or 'uncaring'.

Activity — To be active. To help and give to others may give some relief. However, over-activity is detrimental if it diverts attention from the help you need for yourself.

Reality — Confronting the reality, e.g. attending funerals, inspecting losses, returning to the scene, will all help you to come to terms with the event.

Going — As you allow the disaster more into your mind, there is a need to think about it, to talk about it, and at night to dream about it over and over again. Children play and draw about the event.

Support — It is a relief to receive other people's physical and emotional support. Do not reject it. Sharing with others who have had similar experiences feels good. Barriers can break down and closer relationships develop.

Privacy — In order to deal with feelings, you will find it necessary at times to be alone, or just with family and close friends.

Activity and numbness (blocking of feelings) may be over-used and may delay your healing.

Healing

Remember that the pain of the wound leads to healing. You may even come out wiser and stronger.

Some do's and don'ts

- Don't bottle up feelings. Do express your emotions and let your children share in the grief.
- Don't avoid talking about what happened. Do take every opportunity to review the experience within yourself and with others. Do allow yourself to be part of a group of people who care.
- Don't let your embarrassment stop you giving others the chance to talk.
- Don't expect the memories to go away – the feelings will stay with you for a long time to come.
- Don't forget that your children will experience similar feelings to yourself.
- Do take time out to sleep, rest, think and be with your close family and friends.
- Do express your needs clearly and honestly to family, friends and officials.
- Do try to keep your lives as normal as possible after the acute grief.
- Do let your children talk to you and others about their emotions and express themselves in games and drawings.
- Do send your children back to school and let them keep up with their activities.
- Do drive more carefully. Do be more careful around the home.

WARNING: ACCIDENTS ARE MORE COMMON AFTER SEVERE STRESSES

When to seek professional help

1. If you feel you cannot handle intense feelings or body sensations.
   If you feel that your emotions are not falling into place over a period of time, you feel chronic tension, confusion, emptiness or exhaustion.
   If you continue to have body symptoms.
2. If after a month you continue to feel numb and empty and do not have the appropriate feelings described. If you have to keep active in order not to feel.
3. If you continue to have nightmares and poor sleep.
4. If you have no person or group with whom to share your emotions and you feel the need to do so.
5. If your relationships seem to be suffering badly, or sexual problems develop.

6. If you have accidents.
7. If you continue to smoke, drink or take drugs to excess since the event.
8. If your work performance suffers.
9. If you note that those around you are particularly vulnerable or are not healing satisfactorily.
10. If as a helper you are suffering 'exhaustion'.

Do remember that you are basically the same person that you were before the disaster. Do remember that there is a light at the end of the tunnel. Do remember that, if you suffer too much or too long, help is available.